CW01669296

Brooklands Books

RANGE ROVER

Gold Portfolio

1985-1995

Compiled by R M Clarke

ISBN 1 85520 3200

BROOKLANDS BOOKS LTD.
P.O. BOX 146, COBHAM,
SURREY, KT11 1LG. UK
sales@brooklands-books.com

www.brooklands-books.com

A-RR85GP

Printed in China

Brooklands Books

MOTORING
B.B. ROAD TEST SERIES
Abarth Gold Portfolio 1950-1971
AC Ace & Aceca 1953-1983
Alfa Romeo Giulietta Gold Portfolio 1954-1965
Alfa Romeo Giulia Coupés 1963-1976
Alfa Romeo Giulia Coupés Gold Port. 1963-1976
Alfa Romeo Spider 1966-1990
Alfa Romeo Alfasud 1972-1984
Alfa Romeo Alfetta Gold Portfolio 1972-1987
Alfa Romeo Alfetta GTV6 1980-1986
Alvis Gold Portfolio 1919-1967
AMX & Javelin Muscle Portfolio 1968-1974
Armstrong Siddeley Gold Portfolio 1945-1960
Aston Martin Gold Portfolio 1948-1971
Aston Martin Gold Portfolio 1972-1985
Aston Martin Gold Portfolio 1985-1995
Audi Quattro Gold Portfolio 1980-1991
Audi Quattro Takes On The Competition
Austin A30 & A35 1951-1962
Austin-Healey 100 & 100/6 Gold Port. 1952-1959
Austin-Healey 3000 Ultimate Portfolio 1959-1967
Austin-Healey Sprite Gold Portfolio 1958-1971
Berkeley Sportscars Limited Edition
BMW 6 & 8 Cyl. Cars Limited Edition 1935-1960
BMW 1600 Collection No. 1 1966-1981
BMW 2002 Gold Portfolio 1968-1976
BMW 6 Cylinder Coupés & Saloons Gold P. 1969-1976
BMW 316, 318, 320 (4 cyl.) Gold Port. 1975-1990
BMW 320, 323, 325 (6 cyl.) Gold Port. 1977-1990
BMW 3 Series Gold Portfolio 1991-1997
BMW 5 Series Gold Portfolio 1981-1987
BMW 5 Series Gold Portfolio 1988-1995
BMW 6 Series Gold Portfolio 1976-1989
BMW 7 Series Performance Portfolio 1977-1986
BMW 7 Series Performance Portfolio 1986-1993
BMW 8 Series Limited Edition
BMW Alpina Performance Portfolio 1967-1987
BMW Alpina Performance Portfolio 1988-1998
BMW M Series Gold Portfolio 1976-1997
BMW Z3 & Z3M Limited Edition
Borgward Isabella Limited Edition
Bricklin Gold Portfolio 1974-1975
Bristol Cars Portfolio
Buick Performance Portfolio 1947-1962
Buick Muscle Portfolio 1963-1973
Buick Riviera Performance Portfolio 1963-1978
Cadillac Allanté 1986-1993
Cadillac Automobiles 1949-1959
Cadillac Automobiles 1960-1969
Cadillac Eldorado Performance Portfolio 1967-1978
Checker Limited Edition
Chevrolet 1955-1957
Impala & SS Muscle Portfolio 1958-1972
Corvair Performance Portfolio 1959-1969
El Camino & SS Muscle Portfolio 1959-1987
Chevy II & Nova SS Muscle Portfolio 1962-1974
Chevelle & SS Muscle Portfolio 1964-1972
Caprice Limited Edition 1965-1976
Chevrolet Muscle Cars 1966-1971
Chevy Blazer 1969-1981
Camaro Muscle Portfolio 1967-1973
High Performance Camaros 1982-1988
Camaro Performance Portfolio 1993-2000
Chevrolet Corvette Gold Portfolio 1953-1962
Chevrolet Corvette Sting Ray Gold Port. 1963-1967
Chevrolet Corvette Gold Portfolio 1968-1977
High Performance Corvettes 1983-1989
Chrysler 300 Gold Portfolio 1955-1970
Valiant 1960-1962
Citroen Traction Avant Gold Portfolio 1934-1957
Citroen 2CV Ultimate Portfolio 1948-1990
Citroen DS & ID 1955-1975
Citroen DS & ID Gold Portfolio 1955-1975
Citroen SM Limited Edition Extra 1970-1975
Shelby Cobra Gold Portfolio 1962-1969
Cobras & Cobra Replicas Gold Portfolio 1962-1989
Crosley & Crosley Specials Limited Edition
Cunningham Automobiles 1951-1955
Datsun Roadsters Performance Portfolio 1960-71
Datsun 240Z & 260Z Gold Portfolio 1970-1978
Datsun 280Z & ZX 1975-1983
DeLorean Gold Portfolio 1977-1995
De Soto Limited Edition 1952-1960
Dodge Limited Edition 1949-1959
Dodge Dart Limited Edition Extra 1960-1976
Dodge Muscle Portfolio 1965-1971
Charger Muscle Portfolio 1966-1974
Dodge Viper Performance Portfolio 1990-1998
ERA Gold Portfolio 1934-1994
Ferrari Limited Edition 1947-1957
Ferrari Limited Edition 1958-1963
Ferrari Dino Limited Edition Extra 1965-1974
Ferrari Dino 308 & Mondial Gold Portfolio 1974-1985
Ferrari 328 348 Mondial Ultimate Portfolio 1986-94
Fiat 500 Gold Portfolio 1936-1972
Fiat 600 & 850 Gold Portfolio 1955-1972
Fiat Pininfarina 124 & 2000 Spider 1968-1985
Fiat X1/9 Gold Portfolio 1973-1989
Fiat Abarth Performance Portfolio 1972-1987
Ford Consul, Zephyr, Zodiac Mk. I & II 1950-1962
Ford Zephyr, Zodiac, Executive Mk. III & IV 1962-1971
Ford Cortina 1600E & GT 1967-1970
High Performance Capris Gold Portfolio 1969-1987
Capri Muscle Portfolio 1974-1987
High Performance Fiestas 1979-1991
Ford Escort RS & Mexico Limited Edition 1970-1979
High Performance Escorts Mk. I 1968-1974
High Performance Escorts Mk. II 1975-1980
High Performance Escorts 1980-1985
High Performance Escorts 1985-1990
High Perf. Sierras & Merkurs Gold Port. 1983-1990
Ford Thunderbird Performance Portfolio 1955-1957
Ford Thunderbird Performance Portfolio 1958-1963
Ford Thunderbird Performance Portfolio 1964-1976
Ford Automobiles 1949-1959
Ford Fairlane Performance Portfolio 1955-1970
Ford Ranchero Muscle Portfolio 1957-1979
Edsel Limited Edition 1957-1960
Ford Galaxie & LTD Gold Portfolio 1960-1976
Falcon Performance Portfolio 1960-1970
Ford GT40 Gold Portfolio 1964-1987
Ford Torino Limited Edition 1968-1974
Ford Bronco 4x4 Performance Portfolio 1966-1977
Ford Bronco 1978-1988
Goggomobil Limited Edition
Holden 1948-1962
Honda S500 • S600 • S800 Limited Edition 1962-1970
Honda CRX 1983-1987
International Scout Gold Portfolio 1961-1980
Isetta Gold Portfolio 1953-1964
ISO & Bizzarrini Gold Portfolio 1962-1974
Jaguar and SS Gold Portfolio 1931-1951
Jaguar C-Type & D-Type Gold Portfolio 1951-1960
Jaguar XK120, 140, 150 Gold Portfolio 1948-1960
Jaguar Mk. VII, VIII, IX, X, 420 Gold Port. 1950-1970
Jaguar Mk. 1 & Mk. 2 Gold Portfolio 1955-1969
Jaguar E-Type Gold Portfolio 1961-1971
Jaguar E-Type V-12 1971-1975
Jaguar S-Type & 420 Limited Edition 1963-1968
Jaguar XJ12, XJ5.3, V12 Gold Portfolio 1972-1990
Jaguar XJ6 Series I & II Gold Portfolio 1968-1979
Jaguar XJ6 Series III Perf. Portfolio 1979-1986
Jaguar XJ6 Gold Portfolio 1986-1994
Jaguar XJS Gold Portfolio 1975-1988
Jaguar XJ-S V12 Ultimate Portfolio 1988-1996
Jaguar XK8 Limited Edition
Jeep CJ-5 & CJ-6 1960-1976
Jeep CJ-5 & CJ-7 4x4 Perf. Portfolio 1976-1986
Jeep Wagoneer Performance Portfolio 1963-1991
Jeep J-Series Pickups 1970-1982
Jeepster & Commando Limited Edition 1967-1973
Jeep Cherokee & Comanche Pickups P. P. 1984-91
Jeep Wrangler 4x4 Performance Portfolio 1987-99
Jeep Cherokee & Grand Cherokee 4x4 P. P. 1992-98
Jensen - Healey Limited Edition 1972-1976
Kaiser - Frazer Limited Edition 1946-1955
Lagonda Gold Portfolio 1919-1964
Lancia Aurelia & Flaminia Gold Portfolio 1950-1970
Lancia Fulvia Gold Portfolio 1963-1976
Lancia Beta Gold Portfolio 1972-1984
Lancia Stratos 1972-1985
Lancia Delta & integrale Ultimate Portfolio
Land Rover Series I 1948-1958
Land Rover Series II & IIa 1958-1971
Land Rover Series III 4x4 Perf. Portfolio 1971-1985
Land Rover 90 110 Defender Gold Portfolio 1983-1994
Land Rover Discovery Perf. Port. 1989-2000
Land Rover Story Part One 1948-1971
Fifty Years of Selling Land Rover
Lincoln Gold Portfolio 1949-1960
Lincoln Continental Performance Portfolio 1961-1969
Lincoln Continental 1969-1976
Lotus Sports Racers Portfolio - covering 1951-1965
Lotus Seven Gold Portfolio 1957-1973
Lotus Elite Limited Edition 1957-1964
Lotus Elan Ultimate Portfolio 1962-1974
Lotus Elan & SE 1989-1992
Lotus Europa Gold Portfolio 1966-1975
Lotus Elite & Eclat 1974-1982
Lotus Elise Limited Edition
Marcos Coupés & Spyders Gold Portfolio 1960-1997
Matra Limited Edition 1965-1983
Mazda Miata MX-5 Performance Portfolio 1989-1997
Mazda Miata MX-5 Takes On The Competition
Mazda RX-7 Gold Portfolio 1978-1991
Mercedes 190 & 300 SL 1954-1963
Mercedes G-Wagen 1981-1994
Mercedes S & 600 1965-1972
Mercedes S Class 1972-1979
Mercedes S Class Limited Edition Extra 1980-1991
Mercedes 230 • 250 • 280SL Gold Portfolio 1963-1971
Mercedes SLs & SLCs Gold Portfolio 1971-1989
Mercedes SLs Performance Portfolio 1989-1994
Mercedes 190 Limited Edition Extra 1983-1993
Mercedes CLK & SLK Limited Edition
Mercury Limited Edition 1947-1959
Mercury Comet & Cyclone Limited Edition 1960-1970
Cougar Muscle Portfolio 1967-1973
Messerschmitt Gold Portfolio 1954-1964
MG Gold Portfolio 1929-1939
MG TA & TC Gold Portfolio 1936-1949
MG TD & TF Gold Portfolio 1949-1955
MGA & Twin Cam Gold Portfolio 1955-1962
MG Midget Gold Portfolio 1961-1979
MGB Roadsters 1962-1980
MGB MGC & V8 Gold Portfolio 1962-1980
MGB GT 1965-1980
MGC & MGB GT V8 Limited Edition
MG Y-Type & Magnette ZA/ZB Limited Edition
MGF Limited Edition
Mini Gold Portfolio 1959-1969
Mini Gold Portfolio 1969-1980
Mini Gold Portfolio 1981-1997
High Performance Minis Gold Portfolio 1960-1973
Mini Cooper Gold Portfolio 1961-1971
Mini Moke Gold Portfolio 1964-1994
Morgan Three-Wheeler Gold Portfolio 1910-1952
Morgan Plus 4 & Four 4 Gold Portfolio 1936-1967
Morgan Cars Portfolio 1968-2001
Morris Minor Collection No. 1 1948-1980
Shelby Mustang Muscle Portfolio 1965-1970
Mustang Muscle Portfolio 1967-1973
High Performance Mustang IIs 1974-1978
Mustang 5.0L Muscle Portfolio 1982-1993
Mustang 5.0L Takes On The Competition
Nash & Nash-Healey Limited Edition 1949-1957
Nash-Austin Metropolitan Gold Portfolio 1954-1962
NSU Ro80 Limited Edition
NSX Performance Portfolio 1989-1999
Oldsmobile Automobiles 1955-1963
Oldsmobile Muscle Portfolio 1964-1971
Cutlass & 4-4-2 Muscle Portfolio 1964-1974
Oldsmobile Toronado 1966-1978
Opel GT Gold Portfolio 1968-1973
Opel Manta Limited Edition 1970-1975
Packard Gold Portfolio 1946-1958
Pantera Ultimate Portfolio 1970-1995
Panther Gold Portfolio 1972-1990
Plymouth Limited Edition 1950-1960
Plymouth Fury Limited Edition Extra 1956-1976
Barracuda Muscle Portfolio 1964-1974
Plymouth Muscle Portfolio 1965-1971
Pontiac Limited Edition 1949-1960
Pontiac Tempest & GTO 1961-1965
GTO Muscle Portfolio 1964-1974
Firebird & Trans-Am Muscle Portfolio 1967-1972
Firebird & Trans-Am Muscle Portfolio 1973-1981
High Performance Firebirds 1982-1988
Firebird & Trans Am Performance Portfolio 1993-2000
Pontiac Fiero Performance Portfolio 1984-1988
Porsche 356 Gold Portfolio 1953-1965
Porsche 912 Limited Edition
Porsche 911 1965-1969
Porsche 911 1970-1972
Porsche 911 1973-1977
Porsche 911 SC & Turbo Gold Portfolio 1978-1983
Porsche 911 Carrera & Turbo Gold Port. 1984-1989
Porsche 911 Gold Portfolio 1990-1997
Porsche 911 Takes On The Competition 1990-1997
Porsche 914 Ultimate Portfolio
Porsche 924 Gold Portfolio 1975-1988
Porsche 928 Performance Portfolio 1977-1994
Porsche 928 Takes On The Competition
Porsche 944 Ultimate Portfolio
Porsche 968 Limited Edition Extra
Porsche Boxster Limited Edition
Railton & Brough Superior Gold Portfolio 1933-1950
Range Rover Gold Portfolio 1970-1985
Range Rover Gold Portfolio 1985-1995
Range Rover Takes on the Competition
Reliant Scimitar 1964-1986
Renault Alpine Gold Portfolio 1958-1994
Riley Gold Portfolio 1924-1939
R. R. Silver Cloud & Bentley 'S' Series Gold P. 1955-65
Rolls Royce Silver Shadow Ultimate Portfolio 1965-80
Rolls Royce & Bentley Gold Portfolio 1980-1989
Rolls Royce & Bentley Limited Edition 1990-1997
Rover P4 1949-1959
Rover 3 & 3.5 Litre Gold Portfolio 1958-1973
Rover 2000 & 2200 1963-1977
Saab Sonett Collection No.1 1966-1974
Saab Turbo 1976-1983
Studebaker Gold Portfolio 1947-1966
Studebaker Hawks & Larks 1956-1963
Avanti 1962-1990
Starion & Conquest Performance Portfolio 1982-90
Subaru Impreza Turbo Limited Edition Extra 1994-2001
Suzuki SJ Gold Portfolio 1971-1997
Vitara, Sidekick & Geo Tracker Perf. Port. 1988-1997
Sunbeam Tiger & Alpine Gold Portfolio 1959-1967
Toyota Land Cruiser Gold Portfolio 1956-1987
Toyota Land Cruiser 1988-1997
Toyota MR2 Gold Portfolio 1984-1997
Toyota MR2 Takes On The Competition
Triumph TR2 & TR3 Gold Portfolio 1952-1961
Triumph TR4, TR5, TR250 1961-1968
Triumph TR6 Gold Portfolio 1969-1976
Triumph Herald 1959-1971
Triumph Vitesse 1962-1971
Triumph Spitfire Gold Portfolio 1962-1980
Triumph 2000, 2.5, 2500 1963-1977
Triumph GT6 Gold Portfolio 1966-1974
Triumph Stag Gold Portfolio 1970-1977
Triumph Dolomite Sprint Limited Edition
TVR Gold Portfolio 1959-1986
TVR Performance Portfolio 1986-1994
TVR Performance Portfolio 1995-2000
VW Beetle Gold Portfolio 1935-1967
VW Beetle Gold Portfolio 1968-1991
VW Karmann Ghia 1955-1982
VW Bus, Camper, Van 1954-1967
VW Bus, Camper, Van Perf. Portfolio 1968-1979
VW Bus, Camper, Van 1979-1989
VW Scirocco 1974-1981
Volvo PV444 & PV544 Perf. Portfolio 1945-1965
Volvo 120 Amazon Ultimate Portfolio
Volvo 1800 Gold Portfolio 1960-1973
Volvo 140 & 160 Series Gold Portfolio 1966-1975
Forty Years of Selling Volvo
Westfield Limited Edition

B.B. ROAD & TRACK SERIES
Road & Track on Alfa Romeo 1971-1976
Road & Track on Aston Martin 1962-1990
Road & Track on Audi & Auto Union 1952-1980
Road & Track on Audi & Auto Union 1980-1986
Road & Track on Austin Healey 1953-1970
Road & Track on BMW Cars 1966-1974
Road & Track on BMW Cars 1975-1978
Road & Track on BMW Cars 1979-1983
R & T on Cobra, Shelby & Ford GT40 1962-1992
Road & Track on Corvette 1953-1967
Road & Track on Corvette 1968-1982
Road & Track on Corvette 1982-1986
Road & Track on Corvette 1986-1990
Road & Track on Ferrari 1975-1981
Road & Track on Ferrari 1981-1984
Road & Track on Ferrari 1984-1988
Road & Track on Fiat Sports Cars 1968-1987
Road & Track on Jaguar 1950-1960
Road & Track on Jaguar 1961-1968
Road & Track on Jaguar 1968-1974
Road & Track on Jaguar 1974-1982
Road & Track on Jaguar 1983-1989
Road & Track on Lamborghini 1964-1985
R & T on Mazda RX-7 & MX-5 Miata 1986-1991
Road & Track on Mercedes 1952-1962
Road & Track on Mercedes 1963-1970
Road & Track on Mercedes 1971-1979
Road & Track on MG Sports Cars 1949-1961
Road & Track on MG Sports Cars 1962-1980
R & T on Nissan 300-ZX & Turbo 1984-1989
Road & Track on Pontiac 1960-1983
Road & Track on Porsche 1951-1967
Road & Track on Porsche 1968-1971
Road & Track on Porsche 1972-1975
Road & Track on Porsche 1975-1978
Road & Track on Porsche 1979-1982
Road & Track on Porsche 1985-1988
R & T on Rolls Royce & Bentley 1950-1965
R & T on Rolls Royce & Bentley 1966-1984
Road & Track on Saab 1972-1992
R & T on Toyota Sports & GT Cars 1966-1984
R & T on Triumph Sports Cars 1967-1974
R & T on Triumph Sports Cars 1974-1982
Road & Track on Volkswagen 1951-1968
Road & Track on Volkswagen 1968-1978
Road & Track on Volkswagen 1978-1985
Road & Track on Volvo 1957-1974
Road & Track on Volvo 1977-1994
Road & Track - Henry Manney at Large & Abroad
Road & Track - Best of PS
Road & Track - Peter Egan "At Large"
Road & Track - Peter Egan Side Glances 1983-92
Road & Track - Peter Egan Side Glances 1992-97

B.B. CAR AND DRIVER SERIES
Car and Driver on BMW 1955-1977
Car and Driver on Corvette 1978-1982
Car and Driver on Corvette 1983-1988
C and D on Datsun Z 1600 & 2000 1966-1984
Car and Driver on Ferrari 1955-1962
Car and Driver on Ferrari 1963-1975
Car and Driver on Ferrari 1976-1983
Car and Driver on Mopar 1956-1967
Car and Driver on Mustang 1964-1972
Car and Driver on Pontiac 1961-1975
Car and Driver on Porsche 1955-1962
Car and Driver on Porsche 1963-1970
Car and Driver on Porsche 1970-1976
Car and Driver on Porsche 1977-1981
Car and Driver on Porsche 1982-1986
Car and Driver on Volvo 1955-1986

RACING & THE LAND SPEED RECORD
The Land Speed Record 1898-1919
The Land Speed Record 1920-1929
The Land Speed Record 1930-1939
The Land Speed Record 1940-1962
The Land Speed Record 1963-1999
The Land Speed Record 1898-1999 - Hard Bound
Can-Am Racing 1966-1969
Can-Am Racing 1970-1974
Can-Am Racing Cars 1966-1974
The Carrera Panamericana Mexico - 1950-1954
Le Mans - The Bentley & Alfa Years - 1923-1939
Le Mans - The Jaguar Years - 1949-1957
Le Mans - The Ferrari Years - 1958-1965
Le Mans - The Ford & Matra Years - 1966-1974
Le Mans - The Porsche Years - 1975-1982
Le Mans - The Porsche & Jaguar Years - 1983-91
Le Mans - The Porsche & Peugeot Years - 1992-99
Le Mans - 1923-1999 - Hard Bound
Mille Miglia - The Alfa & Ferrari Years - 1927-1951
Mille Miglia - The Ferrari & Mercedes Years - 1952-57
Targa Florio - The Post War Years - 1948-1973 - H.B.
Targa Florio - The Porsche & Ferrari Years - 1955-1964
Targa Florio - The Porsche Years - 1965-1973

B.B. PRACTICAL CLASSICS SERIES
PC on Austin A40 Restoration
PC on Land Rover Restoration
PC on Midget/Sprite Restoration
PC on MGB Restoration
PC on Sunbeam Rapier Restoration
PC on Triumph Herald/Vitesse

B.B. HOT ROD 'ENGINE' SERIES
Chevy 265 & 283
Chevy 302 & 327
Chevy 348 & 409
Chevy 396 & 427
Chevy 454 thru 512
Chrysler Hemi
Chrysler 273, 318, 340 & 360
Chrysler 361, 383, 400, 413, 426 & 440
Ford 289, 302, Boss 302 & 351W
Ford 351C & Boss 351

B.B. RESTORATION & GUIDE SERIES
BMW 2002 - A Comprehensive Guide
BMW '02 Restoration Guide
Classic Camaro Restoration
Chevrolet High Performance Tips & Techniques
Chevy-GMC Pickup Repair
Engine Swapping Tips & Techniques
Land Rover Restoration Portfolio
MG 'T' Series Restoration Guide
MGA Restoration Guide
Mustang Restoration Tips & Techniques
Practical Gas Flow
The Great Classic Muscle Cars Compared

MOTORCYCLING
B.B. ROAD TEST SERIES
AJS & Matchless Gold Portfolio 1945-1966
BMW Motorcycles Gold Portfolio 1950-1971
BMW Motorcycles Gold Portfolio 1971-1976
BSA Singles Gold Portfolio 1945-1963
BSA Singles Gold Portfolio 1964-1974
BSA Twins A7 & A10 Gold Portfolio 1946-1962
BSA Twins A50 & A65 Gold Portfolio 1962-1973
BSA & Triumph Triples Gold Portfolio 1968-1976
Ducati Gold Portfolio 1974-1978
Ducati Gold Portfolio 1978-1982
Harley-Davidson Sportsters Perf. Port. 1965-1976
Harley-Davidson Super Glide Perf. Port. 1971-1981
Harley-Davidson FXR Series Perf. Port. 1982-1992
Honda CB750 Gold Portfolio 1969-1978
Honda CB500 & 550 Fours Perf. Port. 1971-1977
Honda CB350 & 400 Fours Perf. Port. 1972-1978
Honda Gold Wing Gold Portfolio 1975-1995
Honda CBX 1000 Gold Portfolio 1978-1982
Honda RC30 Performance Portfolio 1988-1992
Kawasaki Z1 900 Performance Portfolio 1972-1977
Kawasaki 500 & 750 Triples Perf. Port. 1969-1976
Kawasaki GPZ 900R Ninja Perf. Port. 1984-1996
Laverda Gold Portfolio 1967-1977
Laverda Performance Portfolio 1978-1988
Laverda Jota Performance Portfolio 1976-1985
Moto Guzzi Gold Portfolio 1949-1973
Moto Guzzi Le Mans Performance Portfolio 1976-89
Moto Morini 3½ & 500 Performance Port. 1974-84
Norton Dominators Performance Portfolio 1949-70
Norton Commando Ultimate Portfolio 1968-1977
Norton Rotaries Limited Edition Extra 1984-1992
Suzuki GT750 Performance Portfolio 1971-1977
Suzuki GS1000 Performance Portfolio 1978-1981
Vincent Gold Portfolio 1945-1980
Yamaha RD350/400 Performance Portfolio 1972-79
Yamaha XS650 Performance Portfolio 1969-1985

B.B. CYCLE WORLD SERIES
Cycle World on BMW 1974-1980
Cycle World on BMW 1981-1986
Cycle World on Ducati 1982-1991
Cycle World on Harley-Davidson 1978-1983
Cycle World on Harley-Davidson 1983-1987
Cycle World on Harley-Davidson 1987-1990
Cycle World on Harley-Davidson 1990-1992
Cycle World on Honda 1962-1967
Cycle World on Honda 1968-1971
Cycle World on Honda 1971-1974
Cycle World on Husqvarna 1966-1976
Cycle World on Husqvarna 1977-1984
Cycle World on Kawasaki 1966-1971
Cycle World on Kawasaki Off-Road Bikes 1972-1979
Cycle World on Kawasaki Street Bikes 1972-1976
Cycle World on Norton 1962-1971
Cycle World on Suzuki 1962-1970
Cycle World on Suzuki Off-Road Bikes 1971-1976
Cycle World on Suzuki Street Bikes 1971-1976
Cycle World on Triumph 1967-1972
Cycle World on Yamaha 1962-1969
Cycle World on Yamaha Off-Road Bikes 1970-1974
Cycle World on Yamaha Street Bikes 1970-1974

MILITARY
VEHICLES & AEROPLANES
Complete WW2 Military Jeep Manual
Dodge WW2 Military Portfolio 1940-1945
German Military Equipment WW2
Hail To The Jeep
Land Rover Military Portfolio
Military & Civilian Amphibians 1940-1990
Off Road Jeeps Civilian & Military 1944-1971
Silhouette Handbook of US Army Air Forces Aeroplanes
US Military Vehicles 1941-1945
US Army Military Vehicles WW2-TM9-2800
VW Kubelwagen Military Portfolio 1940-1990
WW2 Allied Vehicles Military Portfolio 1939-1945
WW2 Jeep Military Portfolio 1941-1945

15/10Z1

Brooklands Books

CONTENTS

ACKNOWLEDGEMENTS

This is the companion volume to our Gold Portfolio covering the Range Rover from 1970 to 1985: the dividing-line between them comes with the introduction of the fuel injected Vogue models.

As this book goes to press, the Range Rover is in its 25th year of continuous production. The new model which will eventually replace it has already been announced, but the old one still commands respect from new car buyers and 4x4 enthusiasts alike - an enviable record indeed.

We are pleased to acknowledge the generous co-operation we have received from the original publishers for the material in this book. Our sincere thanks go to *Australia 4x4, Autocar and Motor, Automobile Magazine, British Car and Bike, Car Australia, Car and Driver, Car South Africa, Carweek, Fast Lane, International Off-Roader, Land Rover Owner, Motor, Motor Trend, Off-Road and Four-Wheel Drive, Overlander, Perfomance Car, Road & Track, Thoroughbred and Classic Cars, What Car?* and *Wheels.*

R M Clarke

The Range Rover was already a 15 year old design when the first article in this book was written. By that stage of their existence, most cars are entering a period of gentle decline but in fact during the next 10 years we see the Range Rover developed as it had never been developed before. Firstly, an injected version of the trusty V8 petrol engine, then a turbodiesel alternative; and over the next few years further engine changes added power (in the shape of the 3.9-litre V8 and a larger turbodiesel). and then economy (in the shape of the Tdi turbodiesel). And all the time, the Range Rover was being pushed gradually further up-market, moving from the executive car alternative of the 1980s to the luxury car alternative of the 1990s.

The impetus for many of these changes came from the USA, where the Range Rover was launched in 1987 (at the ripe old age of 17!) and became a huge success. Demand for more room in the rear and more refinement led to the development of a long-wheelbase version with height-adjustable air suspension, and that in turn pioneered much of the technology for the second-generation Range Rover which was announced in 1994. The well-loved original nevertheless remained in production long enough to see out its 25th anniversary in 1995, rebadged most appropriately as the Range Rover Classic.

As the articles in this collection show, the Range Rover was never without its faults, but its appeal always managed to transcend them. After 25 years in production, it was still widely respected as the best of the luxury off-roaders, and the impact it had on the motoring scene will still be felt long after the last example has left the production lines

James Taylor

VITESSE ENGINE BOOSTS OFF-ROAD KING'S PERFORMANCE

RAPID RANGE ROVER

Don't be too surprised the next time a Range Rover flashes past you on the motorway at over 100 mph. The latest version of that big, luxurious all-terrain car has that capability – along with the necessary suspension modifications to keep it safe at that speed.

You'll recognise the potent newcomer at once as it grows larger in your rear view mirror, since it gets a mild restyling job to go with its increase in power. The most obvious feature is the tough moulded polyurethane front spoiler, wearing a pair of quartz halogen driving lamps, and as the car sweeps by you'll possibly notice the new side rubbing strips.

The secret of the car's livelier performance is fuel injection for the 3.5-litre V8 power plant. The new top specification model, still badged as the Vogue, gets a slightly downtuned version of the engine normally fitted to the Rover Vitesse. That puts out 195 bhp; for the Range Rover output is 165 bhp.

There are transmission improvements, too, the most significant being the introduction of a new four-speed automatic to replace the dated American three-speed "slush pump" which until now was the only box able to cope with the extended high-torque output of the engine used in low ratio, off-road conditions.

The new box is a beefed-up version of the four-speed ZF gearbox as used in BMW and Mercedes Benz models, and retaining the economy-boosting feature of lock-up on the torque convertor when running in top gear.

Fourth gear is actually an overdrive ratio, taking advantage of the car's better power and higher torque output to give relaxed motorway cruising. Since the car still has incredible off-road ability, first gear can be locked in manually for very steep descents.

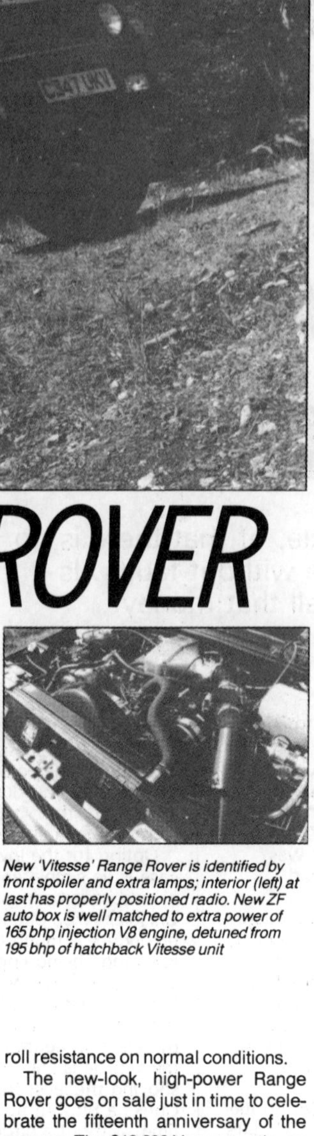

New 'Vitesse' Range Rover is identified by front spoiler and extra lamps; interior (left) at last has properly positioned radio. New ZF auto box is well matched to extra power of 165 bhp injection V8 engine, detuned from 195 bhp of hatchback Vitesse unit

The manual gearbox version of the Vogue uses the same five-speed transmission, but the gearshift linkage is completely revised to give a smoother, slicker change quality.

As the 100mph-plus Range Rover changes lane after overtaking you, you will be struck also by the better control of body roll. That has been achieved without resorting to anti-roll bars, which Land-Rover engineers avoid like the plague in order to ensure that the Range Rover retains its excellent off-road capability. The answer in the new Vogue has been to fit dual-rate springs and bigger bushes along with revised damper settings to increase roll resistance on normal conditions.

The new-look, high-power Range Rover goes on sale just in time to celebrate the fifteenth anniversary of the marque. The £18,696 Vogue continues to increase in luxury appeal because that is definitely where the demand lies; in the past year more than 80 per cent of total Range Rovers sold were Vogues.

Hence the power boost brings with it an improvement in the interior refinement of the big machine. For a start, the new ZF gearbox allows a narrower transmission tunnel, giving more front legroom.

The radio, originally sited awkwardly to the driver's right, has been moved to a more accessible – and more normal – position in the centre.

Interior looks are enhanced by new trim colour options like Silver Grey or Bracken, keyed to the body colour. Here, too, there have been notable advances, including one-shot paint finishing of the entire body with cathodic electro coat priming and clear-over-base metallic finishes.

Better looks alone might have sufficed, had it not been for the obvious interest shown by many owners in getting more performance out of their Range Rovers, with turbocharger conversions – or even complete engine transplants – becoming ever more popular. Range Rover engineers boast that their own improvement gets the car back in there with the general flow of today's traffic. While the claimed 107 mph top speed might be a little over the top in terms of legality in Britain, the 2.5 seconds chipped off the 0-60 mph acceleration time is a soundly practical feature. To match the higher speed potential there are new tyres available, too – specially developed by Michelin XM + S 200 in 205 × 16 size or Avon Rangemaster 215/75 × 16s.

Climb into the new Vogue and you'll notice some additional improvements, starting with the puddle lamps in the base of the doors, which now also have red warning lamps when opened. On the manual version the gear lever is a neater, shorter gear lever, falling much more naturally to hand. The handbrake lever, too, is more accessible and easier to use.

Electronic control of the Lucas L fuel injection system ensures quick starting and good efficiency in all conditions. The power boost brings with it an alteration in the engine's torque characteristics, and certainly it is noticeably "peakier" when driven hard on the road with string pulling power coming in only above 3000 rpm.

Peak torque comes at a rather high engine speed of 3200 rpm, and even though the total amount of torque on tap is a hefty 206 lb ft the engine is patently not ideally suited to hauling its way trudgingly through slippery mud. The sales men from Solihull, however, say they can live with that, since few Range Rover Vogue owners will want to put their car's ultimate ability to the test in really tough cross-country conditions. The demonstration course they had laid out included farmyards and tracks through fields, with a few mild yumps and gullies thrown in – nothing really demanding, but quite up to what an average owner might expect his car to cope with. The Range Rover personnel aren't even trying to accentuate the car's off-road role. They say they see it competing with other luxury estates – a straight fight against the big Mercedes and Volvos.

MODEL: RANGE ROVER VOGUE	
DATE ON SALE:	NOVEMBER '85
RANGE:	4-door estate
ENGINE:	3528cc V8, 165 bhp
PERFORMANCE:	Max 107 mph
MPG:	15
PRICE:	£18,696

Range Rover Vogue Automatic

A magnificent vehicle, of that there is no doubt, but not one without faults. Is it really worth all that money?

After our all too brief introduction to the latest Range Rover Vogue, reported in the December issue, a comprehensive *Off Road and 4 Wheel Drive* test was obviously called for. Our test vehicle was a Vogue fitted with the optical ZF four-speed automatic transmission. It was finished in Astral Silver and looked magnificent.

The engineering concept of the Range Rover is, and always has been, absolutely right and, over the 15 years or so of its life, a degree of development work has kept it ahead of the competition. More recently, a determined effort has been made to give it that final polish, to push it inexorably towards the top of the luxury car sector.

The latest model sports the 165bhp fuel injected version of the ubiquitous 3½ litre V8. In this form, detuned from its Rover SD1 Vitesse specification, it produces a massive 206lb/ft of torque that is more than adequate to propel this heavy car rapidly to an official top speed of 107mph.

Ratio changes in the delightfully smooth ZF four-speed automatic transmission are barely discernable; only a change to the exhaust note and a slight surge when under full power give the game away, if you really concentrate. Under normal driving conditions, both on and off the road, it is virtually impossible to catch this 'box out. It is always in the correct gear and this helps enormously to promote confidence in the vehicle's controlability, unlike some automatics. It is possible, however, to provoke a degree of indecision in the gearbox by approaching an incline on a very light throttle opening. As road speed lessens, some hesitation is detectable until the system decides that you don't know what you are doing either; it then makes its selection, and correctly too.

Opting for the low range of the transfer case is as simple and straightforward as it *should* have been with the original Chrysler A727 three-speed auto box. The procedure is much the same, but slicker, low range being engaged when either stationary – by merely moving the lever firmly forward – or on the move at no more than 5mph or so. This is achieved by selecting neutral on the auto box while the vehicle is still rolling and rapidly shoving the lever from high to low. It clunks a bit, but at least it goes in, first time, without a struggle.

The high torque and incredible flexibility of the V8 power unit means that only rarely does one require the use of low box. Some Range Rover owners will never take advantage of the facility in earnest, though curiosity may tempt them into dabbling with it just to see if it is still operating. For *precise* manoeuvring of a heavy trailer or caravan, for recovery of larger vehicles and for crawling over severe cross country conditions, it is, of course, essential.

With the added advantage of the torque converter's multiplying effect, the automatic transmission is even more versatile, particularly as the ratio changes can be limited by placing the selector in the appropriate gear. Thus, locked in first, the car will pull away with the minimum of fuss, without wheelspin and on a surface that would require the greatest delicacy and experience in a machine equipped with the manual gearbox. The only time an automatic transmission is found wanting in off-road conditions is on extremely steep descents, but it does have to be very steep before it becomes alarming.

Range Rovers have, over the years, achieved the reputation of cornering with a certain degree of nautical style. These 'leanings' led to a great deal of criticism from those users who didn't perhaps need or appreciate the vehicles cross country ability, to which the highly flexible suspension is a major contribution. Basically, if you desire a 'jack of all trades' then you have to accept some design compromise. A great many Range Rovers have been modified to improve their roadholding by the addition of after market anti-roll bar kits. There is no doubt that they work but not without sacrificing some of the vertical wheel travel and not enough to have any serious effect on the off-road capability.

Now, Land Rover Ltd have come up with their own solution to the problem, and not before time either. What they have done is to fit dual rate coil springs at the rear end, larger radius arm bushes at the front and revise the damper settings all round. The result is vastly improved roll resistance, giving a flatter cornering stance, without any detrimental effect on axle articulation. Quite a technical achievement that; the vehicle has

Range Rover Vogue Automatic

highway performance, handling and ride quality akin to the best luxury cars, yet is second only to its own stablemate, the Land Rover Ninety V8, in the off-road stakes.

Two weeks living and playing with the new Vogue was ample to reveal a disconcerting number of undesirable features, however. Some were due to basic design, but most were associated with poor quality trim. Now, remember, ownership of this magnificent *looking* machine will set you back £19,000. That fact alone kind of puts things in perspective. People who can afford to pay that sort of price are usually very keen on getting value for their money. They are not going to be at all impressed with paint that is so soft that the slightest brush with vegetation in a country lane produces great scratches along the vehicle's flanks. While they will surely be mightily impressed by the stereo radio and cassette player, its incoming signal is via one of the cheapest and nastiest looking, manually operated, telescopic aerials one could hope to find on a local market stall.

The fuse box, sensibly located in the fascia behind a concealing panel, includes a pair of tongs and a selection of spare 'spade' fuses. Access is by sliding the aforementioned panel to the left and lifting it out. Unfortunately the four plastic retaining hooks are not particularly strong and, with two broken, the panel is reluctant to stay in place. We're not finished yet. The ashtray for the front seat occupants is mounted in the centre console, immediately ahead of the gear selector. Not a bad place to put it you might think, but, when 'Park' is selected, it is all to easy to stub out one's cigarette on the knob of the change lever as it lurks right above the ashtray. When the time comes to empty the accumulated fag ends etc, you will find that removing the ashtray from its housing is not so simple. In the test vehicle the whole trim panel came away from the console. Of interest is what was revealed below, a piece of thick plywood packing under the change lever assembly. No comment! Another piece of trim that persisted in falling off was the strip of grained plastic that decorated the inside of the right-hand front door pillar by the windscreen.

There is an excellent indication of how quiet the Vogue is on the road, No, you can't hear the clock ticking, but the speedometer does and very irritating it is too, clicking quietly away to itself, drowned out only by the rush of wind noise from about 55mph upwards.

For sheer unadulterated comfort the Range Rover's seats take some beating. They have always been good, but they are now better than ever. The front inertia belts have a movable top strong point, so they can be set to the most comfortable fit for any sized occupant. Both front seats are fully adjustable and it pays to spend a little time to get them absolutely right. With fold-down armrests and a fully automatic transmission, this is a very 'laid back' motor car to drive. Actually it could be even more comfortable if the wooden door cappings were replaced with some form of padding too.

Colour page, main pic: **The Range Rover's off-road ability is legend and the Vogue sets new standards.** *Inset:* **Only the nudge bar distinguishes the Janspeed twin turbo version from the standard Vogue.** *Above:* **The Range Rover's seats take some beating for sheer comfort.** *Left:* **Still 'top of the heap', the Vogue crosses arduous terrain with little effort.**

Passenger comfort in the rear is also very good, there being only one complaint – the inertia belts fitted to the outside seats. There is a tendency for the strap to ride up on the occupant's neck, not seriously wtih most adults, but definitely a problem for children. Still, one should be grateful for small mercies; at least rear seat belts are fitted.

The cavernous cargo area is thickly carpetted, in keeping with the rest of the vehicle and this includes the load bearing surface of the tailgate itself. The untidy bits, like the spare wheel, are neatly boxed-in and a removable, folding, parcel shelf is installed to secure your valuables from the eagle eyes of the light fingered brigade. The latter, made from some kind of pressed fibre board, should not be taken too seriously.

It is something of a mystery why Land Rover still persist in not offering a single piece lift-up tailgate, as a no-cost option, in lieu of the horizontally split version. A number of companies already produce these for their various conversions and a particularly good one is made by Alan Smith Garages of Derby as an after-market fitting. While acknowledging that the split gate is more versatile, a lift-up hatch is much more convenient. The main drawback of the split gate is the problem it can pose when unloading small items from the back. If you reach in through the top section you immediately get mud and grime all over your clothes. (The back end of a Range Rover is alway dirty!) If you then decide to drop the bottom half to avoid this, you will not only get your hands grubby, but you won't be able to reach in far enough to get what you are after anyway. The next stage is an undignified clamber into the rear end . . . Come on Solihull, where is your sense of adventure? At least *offer* a single piece tailgate.

While you are about it, you can re-position the washer jet on the rear screen, 'cos it only sprays down one side of the wiper blade, the water being obstructed by the arm as it sweeps back up the glass. It has been that way since the rear washer was introduced in the early '70s. Yes, it works, but is it really good enough on a vehicle costing £19,000 or more?

Externally the Range Rover's shape has changed not at all since it was introduced. Extra holes were cut in the side to accommodate four doors instead of two, but other than that, the classic and award winning bodywork soldiers on. To identify the new Vogue, however, the addition of some plastic trim was felt necessary and this takes the form of a black rubbing strip, incorporating a silver style line, that runs down both sides of the car. Under the front bumper is mounted a substantial spoiler moulded in black polyurethane. At first glance any true off-road enthusiast will throw up his hands in horror. We certainly had our doubts about it. In the event, however, it transpires that the spoiler does not cause any problems when travelling across country as it is tucked well back behind the line of the approach angle. Should you be careless enough to ram a tussock or two then it is flexible and forgiving enough to let you get away with it.

Housed within this structure are two quartz-halogen driving lights which vastly improve the available illumination and are more than adequate for the performance of the vehicle. It was not possible to assess the effect of the spoiler when wading, but presumably, at the appropriate depth, it would create a hefty bow wave. This could be an advantage, provided that one keeps moving, as a hollow forms in the water behind the

Technical Specification

Engine(Injected)	
Type	All aluminium water cooled V8 with overhead valves
Bore	88.9 mm
Stroke	71.1 mm
Capacity	3528 cc
Compression Ratio	9.35:1
Maximum Power	165 bhp (123 Kw) at 4750 rpm DIN 70020
Maximum Torque	206 lbf. ft. (280 Nm) at 3200 rpm DIN 70020
Fuel Injection	Lucas 'L' electronic fuel injection
Engine (Carburettor)	
Maximum Power	127 bhp (94 Kw) at 4000 rpm DIN 70020
Maximum Torque	194 lbf. ft. (263 Nm) at 2500 rpm DIN 70020
Carburettors	Twin Zenith Stromberg
Gearbox	
Type	Land Rover LT77 five-speed constant mesh with five forward and one reverse ratio, oil feed pump and filter.
Ratios	Fifth 0.77:1
	Fourth 1.0:1
	Third 1.39:1
	Second 2.13:1
	First 3.32:1
	Reverse 3.42:1
Automatic	
Type	ZF4 HP22 four speed automatic with lock-up clutch by-passing torque converter in top gear
Ratios	Fourth 0.72:1
	Third 1.0:1
	Second 1.47:1
	First 2.47:1
	Reverse 2.08:1
Transfer Gearbox (all models)	
Type	Land Rover LT230 two speed transfer gearbox featuring permanent four wheel drive with lockable centre differential
Ratios	High 1.19:1
	Low 3.32:1
Tyres	
Vogue	Michelin XM+S 200 205×16 or Avon Rangemaster 215/75×16
Two/Four Door	Michelin XM+S (as previously fitted)
Price	
Two door	£15,164.28
Four door	£16,251.89
Vogue	£18,696.22
ZF Automatic Transmission	£997,91

'wave' and allows the engine to run happily in considerably less depth than the rest of the car. Hopefully the driving lights are well sealed against water ingress.

Aesthetically the spoiler is quite pleasing as it successfully covers up the lumpy bits underneath and gets rid of the long legged look of old. It can only improve the air flow around the far-from-aerodynamic front end.

Part of the overall trim package for all new Range Rovers is the inclusion of wrap-round plastic end caps on both front and rear bumpers. They look reasonably neat and tidy, but it was immediately apparent that, when off-roading, the ever present passing vegetation was going to get hooked up in them, and so it proved. It was not done deliberately, but, on returning from the first foray, a piece of shrubbery was found wedged tight into the corner of the bumper. In getting there it had fractured a flimsy plastic mounting bracket attaching the end cap to the bodywork and severely strained the screw threads (plastic again) into which the two main securing bolts fastened. Sorry Land Rover, but you really are going to have to do a lot better than this. It is accepted that the 'Hoorays' are not likely to venture off the tarmac, but the real 'Huntin', Shootin' and Fishin'' set do and they will not be impressed by their inadvertent littering of the countryside with bits of expensive plastic.

Driving the new fuel injected Vogue is an experience that is not to be missed and, with the specially developed ZF4 HP22 automatic transmission, it is incredibly enjoyable too. The engine fires up instantly, (there is no choke) and pulls well from cold. The Lucas fuel injection system is electronically managed with constant monitoring to ensure optimum efficiency. As a further aid to economy the fuel supply is cut off completely when the engine is 'overrunning' – no point in wasting the stuff.

Pulling away from rest can be as quiet as you like or downright exciting. Firm acceleration produces a rapid response, but the vehicle remains taut and utterly controllable at all times. Ratio changes are made without a trace of a jerk, merely an alteration of the exhaust note and a mild surge of power. Until the novelty of driving a 'quick' Range Rover wears off it is difficult to burble around the country lanes at a sensible speed, hence the petrol consumption was pretty horrific. Including a long blast down the M4 to Bridgend, the first week showed a return of only 14mpg. With sanity restored and a determined effort to take it easy, this figure was later pushed up to a maximum of 18mpg.

This represents a fairly reasonable thirst for a big engine doing a lot of work and includes rather more off-road activity than most Vogues will experience. Perhaps the most telling way to quote fuel costs is in terms of pence per mile. For the duration of the test period this amounted to approximately 13p. Doesn't sound a lot does it? Think about it and remember, that is for fuel alone. Where, oh where is that diesel Range Rover?

The vehicle's off-road ability is already a legend. Suffice to say that it is still at the 'top of the heap' when it comes to this facet of its character. Endless torque and the facility to apply it smoothly via the ZF box ensures that the Vogue can cross arduous terrain with little or no apparent effort. The punishment meted out by rough and rutted tracks is soaked up with ease by the supple coil suspension and a surprisingly high average speed can be maintained without discomfort to the occupants.

The Range Rover is, without a shadow of doubt, a magnificent vehicle. The engineering concept, honed to its present state, is superb. The new Vogue model is a brilliant combination of power and performance with an on-road ride quality that is easily on a par with the very best. Off-road, it sets new standards yet again.

However, all these superlatives can only be applied to the underlying machine. The add-on cosmetic bits and pieces, those items that generally have the casual onlooker drooling at the apparently extravagant luxury of it all, just do not measure up. It is, after all, still intended as a dual purpose vehicle, albeit at the top end of the market; £19,000, remember? It is no longer an honest, value for money, machine and hasn't been for some time now. In fact, the *car* is probably even more of a poseur nowadays than many of the people who drive it!

Sometime in the new year, Land Rover Ltd is re-entering the North American market with the Range Rover. It will be a worrying time for them. Perhaps they should have some profound talks with Jaguar Cars before they go?

Full blown version

Driving Janspeed's turbo Vogue

While the Range Rover Vogue is now endowed with the sort of power and performance to complement its sophisticated up-market image, it never used to be. Early versions still retained the somewhat strangled 126bhp V8 with its sub-3000rpm torque churning capabilities contributing to that legendary off-road performance.

Ideal mud-plugging engine characteristics make for less that ideal high speed cruising, and Wellingtons are arguably more often found pushing Range Rover throttles to the floor on motorways, than squelching through fields of mud.

Add to that vertical brick aerodynamics, and a Range Rover needs all the grunt it can get to maintain reasonable road averages – hence, the latest 165bhp Vogue. But, short of slotting that fuel injection engine in an earlier Vogue, owners of earlier versions have few options available to increase on-road punch.

Performance specialists Janspeed know this. Janspeed also know a lot about Range Rovers, being the official Land Rover diesel conversion people. The firm has a thriving business of pulling V8s out of Range Rovers and slotting the 2.4 litre turbocharged VM diesel engine in place, producing a car with excellent off-road abilities which is perhaps best suited to countries where the economic advantages of diesel engines are greater than those in the UK.

But, with a maximum power of 100bhp at 4200rpm the turbo diesel Range Rover conversion takes the car even further away from the up-market Vogue high-speed cruising brief.

That's why Janspeed has engineered turbocharger conversions onto the carburettor V8. The first was the archaic and troublesome suck-through system, but that was soon replaced by the currently ubiquitous blow-through type. Janspeed offers a variety of conversions, from a single Garrett TO4 turbo, with or without an intercooler, to a twin Roto Master RM60 set-up, again intercooled or not.

The twin system will only fit RHD Range Rovers, and although the performance improvement of either single or twin set-ups is about the same, Janspeed considers that the slightly swifter throttle response inherent from the two smaller Roto Masters, with their lesser inertia in rotation, means that this is the ideal set-up for use with an automatic transmission.

Excluding VAT, the top-line turbo conversion – twin turbo with intercooler – costs £1740, plus £380 for fitting. The cheapest is the single Garrett with intercooler at £1160 with the same fitting cost.

Intercooled twin turbo

In keeping with the up-market Vogue perspective, we decided to try an 'original' carburettor automatic Vogue fitted with Janspeed's top-line intercooled twin turbo kit.

Immediately on opening the bonnet it was apparent that this was no aftermarket bolt-on bodge. An attractive cast, baffled, plenum tube dominated the compartment which feeds the compressed charge via flexible silicone tubing into the two standard Stromberg carburettors. Modular wastegates control the inlet manifold pressure to 5psi, and the power unit is kept under control by a Micro Dynamics engine management system. In keeping with the high speed touring brief, Janspeed's own intercooler sits up front below the engine coolant radiator, and as in all Janspeed turbocharger installations, an engine oil cooler is used to keep the lubricant in its best temperature range.

The claimed result is in excess of 220bhp and a top speed over 100mph with 60mph coming up from rest in 12sec, and with engine management sohistication and serious cooling attention it's claimed that fuel consumption will remain 'very much the same as standard'.

'Our' car was also fitted with £98 worth of Janspeed rear anti-roll bar and an imposing £125 front nudge bar. It was also an automatic – an early one – the 3-speed Chrysler unit.

Driving the Janspeed car immediately after the latest 4-speed automatic Vogue first indicated just how good that later ZF 4-speed is. Not an excessively sluggish old dog in itself, in comparison the Chrysler unit nevertheless serves to underline the significant strides which have taken place in automatic transmission design over recent years.

But we were't testing the transmission! How good was the turbocharged engine to drive, and how well does the package suit the Range Rover? Janspeed reckons that, with its new system, turbo lag has been 'completely eliminated' . . . well it is hardly perceptible. But, it's easy to see why the firm recommends the twin turbo conversion to be used with automatic transmissions. Never ideal partners, the off-boost lack of urge inherent in a turbocharged engine, mitigates against the right sort of torque response characteristics being fed to an automatic transmission for it to respond accordingly and smoothly.

Most turbo/automatic combinations require somewhat unnatural throttle control if smooth power deliveries are to be maintained to the driving wheels. But in a Range Rover equipped with a professional turbocharger conversion this effect is minimised, and it doesn't take long for a driver to become familiar with the required control characteristics. Janspeed's conversion pumps out no less than 250lb.ft of torque at 2700rpm . . . on boost, and that's the important bit.

That sort of available power is an ideal characteristic for a comparatively heavy 4WD off-roader. Connected to a manual gearbox it's certainly likely that this near lag-free turbo power house is a particularly attractive complement to the Range Rover. Connected to an automatic transmission, however, it's *almost* as useful – if you can put up with neck-jerking full-torque power changes up the box, which can not only be painful to occupants, but sound as if they're hardly prolonging the active life of the transmission either.

Janspeed's anti-roll bar does its job well, and the earlier car seems to approach the controllability and lean angles of the better strapped-down later version. A worthwhile addition.

And this professionally engineered and installed turbocharger conversion is equally worthwhile as the ideal way to increase performance, particularly if it has a manual gearbox.

RANGE ROVE

Upmarket oiler, or farmer's friend? Expensive to buy but cheap to f
Solihull's world-beater has at last been fitted with a diesel engine and the combina

Turbo Diesel

nearly as fast as the petrol model

Development of Britain's world-beating, fashionable off-roader has been lethargic to say the least. It has taken well over a decade for such niceties as four doors, five speeds and automatic transmission to appear, and yet while the opposition has been able to boast all this and more, none has so far managed to undermine Solihull's supremacy.

Running a Range Rover has never been cheap, however. The latest carburetted version of the lusty V8 consumes two-star at the rate of 15 mpg, but arguably – in this country at least – those who can afford the necessary £16,886 to own the basic five-door, five-speed version can support the engine's thirst. Europeans are less tolerant, however, and in countries such as Italy, where the diesel engine does not suffer from a commercial-vehicle-only stigma, fuel at half price and an engine which uses less of it into the bargain are persuasive arguments for oil-burning motivation – even with a basic £18,109 price tag.

Thus the news that Land Rover expect the newly announced turbo-diesel-engined Range Rover to account for 60 per cent of the model's sales on the Italian market comes as no surprise. Over here, the model's launch has been slightly lower key. The manufacturers are anxious not to compromise the Range Rover's upmarket image, and have concentrated more on the efforts of their engineers to isolate the compression ignition engine's coarser manners, and keep its performance on a par with its petrol-engined sister, rather than make too much capital out of the considerable improvement in fuel consumption.

Appropriately, perhaps, the diesel engine in the Range Rover is made in Italy at the VM factory in Bologna. Designated HR 492HI, the unit displaces 2393 cc from four 92 x 90 mm cylinders with a compression ratio of 22:1. The cylinder heads are four sep-

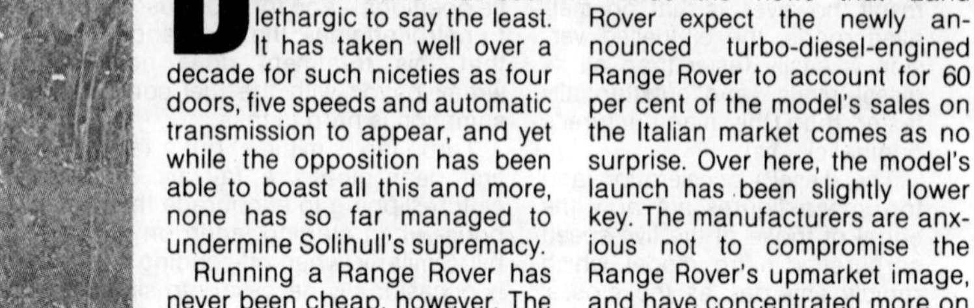

arate aluminium alloy items – simply because VM make a family of engines including five- and six-cylinder versions, and this saves the need for different castings. The basic block lives on from the original turbo diesel Rover SD1 model.

Fuel feed is via a KKK turbocharger with integral wastegate, and the relatively high boost pressure of over 8.2 bar (12 psi) is delivered through an air-to-air intercooler sited to the offside of a very large radiator, which also has the thermostatically controlled engine oil cooling within its dimensions. The engine produces 112 bhp at 4200 rpm and a formidable 183 lb ft of torque at only 2400 rpm, both figures close to the carburetted petrol V8's 125 bhp/4000 rpm and 190 lb ft/2500 rpm delivery.

There has been some criticism regarding the first diesel Range Rover's performance. Since the launch, Land Rover have shortened the overall final drive; they say this was done simply to make the transfer gears quieter. The result, however, is that on metalled roads, the oil-fuelled version is easily faster than all its diesel rivals (and substantially faster than the manufacturer's original claims).

The diesel's acceleration and top speed figures are now the equal of those of the five-speed carburetted petrol model, which rapidly emerges as the closest opposition. The lower gearing gives a 25.1 mph/1000 rpm top gear against the 25.8 mph/1000 of the petrol (and the first diesel models). Now, the top speed of 95.8 mph is all but identical to the V8, as is the zero to 60 mph acceleration time of 14.5 sec (compared with 14.6 sec). The all-important 30-50 mph fourth gear increment is 9.6 sec for the diesel, 9.5 sec for the petrol. We have not tested all the diesel-engined rivals, but the Range Rover's 112 bhp and 183 lb ft of torque powering its 1938 kg mass suggests a stronger combination than the 3980 cc Toyota Landcruiser's 99 bhp and 171 lb ft (1971 kg), which comes closest, or the 2346 cc turbo Mitsubishi Shogun with its 84 bhp and 129 lb ft of torque.

Although the on-paper figures, achieved under traffic-free testing conditions, show the two-tonne diesel machine to be a more than respectable performer, the *way* in which it performs deserves some comment. It is no longer possible to be lazy with the gearbox as you can with the petrol version, and still make reasonable progress. The turbo motor has a comparatively narrow power band, from a little above 2000 rpm when the boost starts in earnest up to 4800 rpm when the governor cuts in. Keeping the machine moving briskly is simply a matter of making sure that you are in the right gear before making a manoeuvre.

At lower speeds it is sometimes necessary to drop two gears in order to overtake energetically, while on the motorway, the driver must be ready to drop out of top gear to stay with the traffic on long inclines. Subjectively, this is not as much work as it sounds and to the driver used to a diesel engine it's second nature. Once the essential 2000 rpm-plus is showing, however, the KKK turbo spins easily with a traditional muted turbo whistle. The engine's response is then immediate and the surge of midrange power extremely impressive. To make brisk progress it is essential always to keep your foot flat on the floor – the diesel does not respond sensitively to part throttle openings – and for those used to petrol engines, the assurance that this treatment does not wreak havoc with the fuel consumption is hard to take.

Turbo lag is minimal but a tall first gear means a fair bit of clutch-slipping to encourage the boost when starting laden on a hill. Similarly, when off-roading it is occasionally necessary to slip the clutch to spin up the turbo in awkward situations. Once again it's a simple matter of technique.

The five-speed gearbox, originally lifted in 1983 from the Rover saloon, is similar to that fitted in the petrol machine, although the clutch is smaller with softer springs to suit the oiler's different power characteristics. The shift is light and the action short, although on the test car, the reverse gate was not clearly defined, and it was possible to confuse first and reverse. Now long gone is the legacy of the whining, clonky transfer box that characterised the first 10 years of Range Rover production, and the example in the diesel is yet further refined, with post-launch revisions to both input and output gear design, in order to reduce noise along with the shorter overall gearing.

The chassis is unaltered save for slightly stiffer front springs to cope with the extra weight of the diesel and although understeer and body roll are fairly pronounced on tarmac, the commanding view afforded by the exalted seating position allows speedy and safe progress once you have become used to the machine's dimensions. The steering is accurate, if low-geared, and inputs are quickly translated into body roll, which gives the driver a good feel for the car.

Off-road the machine's abilities are already well chronicled and the limit is usually defined by the driver's skill, courage and physical fitness. Often in that precise order.

Remarkably, the metalled road ride is firm yet comfortable, and the cabin has a reassuringly stiff feel to it at all times. Brakes are disc all round, as before, and are powerful and nicely weighted.

Unsurprisingly, the four-cylinder diesel lacks the low speed refinement of the petrol V8. It is also noticeably less smooth than the five-cylinder engine in the Mercedes 300 G-Wagen, but in absolute terms the VM four's diesel character is reasonably well suppressed, becoming relatively more so as the revs rise. At tickover, passers-by are in no doubt as to the nature of the underbonnet power, and there is a distinctive diesel induction roar as the accelerator is released on the overrun. The noise meter confirms the subjective impressions however, and at a constant 70 mph and under full throttle in second gear the diesel, equipp with its modified transfer box now quieter than the previc petrol model (77/80 db aga 78/83 db respectively).

Otherwise, the Range Rove as before and seats and inte are similar to those of the fo door Vogue petrol. Included standard on the diesel is cen locking, but initially at least, turbo D will only be availa here as a four-door, with a h price tag of £18,109, wh makes it easily the most exp sive of all barring the £19, Range Rover Vogue. Mercedes 300GD is priced ne by at £17,240 with its sup smooth five-cylinder diesel

Above: Seats are but well-shaped, g commanding view

Left: Latest RR fa have clear instrum and convenient switchgear

VM turbo-diesel weighs more than V8 but fits under bonnet with room spare

ine and transmission, but it is astly slower and smaller physically. The top of the range Shogun estate at £13,799 seems a eal bargain by comparison, but gain is slightly smaller and ower, while unashamedly pitching itself a little further ownmarket in terms of equipment, accommodation and canet aspiration. None of the iesel-engined rivals, including he Nissan Patrol 3.3 five speed 13,196) and the Toyota andcruiser (£14,649), can atch the Solihull newcomer's peed or 27.4 mpg touring fuel onsumption. We suspect that he Rover's 22.4 mpg overall figre would be hard to beat too, nd in view of the performance, a 9 per cent improvement over its etrol sister is nothing less than xceptional.

Land Rover expect to sell bout 300 of the new model in his country next year, mainly to uck operators and farmers who ave bulk supplies of Derv. The rice of diesel Range Rover style high, but then the company onfidently assert that depreciaon will be less. Whether this iesel's undoubted capabilities, llied with its frugal fuel conumption, can overcome the stiga of an accompanying taxi ratle when parking outside the pera remains to be seen. Whether potential owners will be pera-goers at all is another atter. It seems more likely that and Rover have simply plugged he remaining gap in their range ith a diesel alternative to the erman and Japanese opposion, although it does seem a little uzzling that there is no basic wo-door stripped-out version on ffer at a price closer to the vals. This is apparently a future ossibility, dependent on uccess.

Solihull will be watching the arket closely, and the irony of he situation is that to gain aceptance, "Europe's fastest ual-purpose diesel" must break nto the unique cachet that the arque has itself created. M

MOTOR ROAD TEST
LAND ROVER RANGE ROVER TURBO D

PERFORMANCE

WEATHER CONDITIONS

Wind	10 mph
Temperature	63 deg F/ 17 deg C
Barometer	29.8 in Hg/1011 mbar
Surface	Dry tarmacadam

MAXIMUM SPEEDS

	mph	kph
Banked Circuit (5th gear)	95.8	154.1
Best ¼ mile (5th gear)	97.0	156.1
Terminal speeds:		
at ¼ mile	68	109.4
at kilometre	84	135.2
Speeds in gears (at 4500 rpm):		
1st	26.2	42.1
2nd	40.8	65.8
3rd	62.6	100.7
4th	87.0	140.0

ACCELERATION FROM REST

mph	sec	kph	sec
0-30	4.2	0-40	3.1
0-40	6.7	0-60	6.0
0-50	10.1	0-80	10.0
0-60	14.5	0-100	15.8
0-70	20.4	0-120	24.1
0-80	29.5	0-140	44.4
Stand'g ¼	19.7	Stdg km	36.7

ACCELERATION IN TOP

mph	sec	kph	sec
20-40	19.6	40-60	11.7
30-50	14.5	60-80	8.6
40-60	14.7	80-100	9.6
50-70	17.1	100-120	12.0
60-80	22.2		

ACCELERATION IN 4TH

mph	sec	kph	sec
20-40	11.5	40-60	7.0
30-50	9.6	60-80	6.0
40-60	10.5	80-100	7.2
50-70	12.2	100-120	9.4
60-80	16.0		

FUEL CONSUMPTION

Overall	22.4 mpg 12.6 litres/100 km
Touring*	27.4 mpg 10.3 litres/100 km
Govt tests	25.5 mpg (urban) 34.1 mpg (56 mph) 24.4 mpg (75 mph)
Fuel grade	Derv
Tank capacity	81.8 litres 18 galls
Max range*	493.2 miles 793.6 km
Test distance	702 miles 1130 km

*Based on official fuel economy figures – 50 per cent of urban cycle, plus 25 per cent of each of 56/75 mph consumptions.

STEERING

Turning circle	11 m 36 ft
Lock to lock	3.4 turns

NOISE

	dBA
30 mph	69
50 mph	74
70 mph	77
Maximum†	80

†Peak noise level under full-throttle acceleration in 2nd

SPEEDOMETER (MPH)

True mph	30	40	50	60	70	80	90
Speedo	32	42	52	62	72	82	90

Distance recorder: 1.2 per cent slow

WEIGHT

	Kg	cwt
Unladen weight*	1938	38.2
Weight as tested	2088	41.1

*No fuel

Performance tests carried out by *Motor's* staff at the Motor Industry Research Association proving ground, Lindley, and Millbrook proving ground, near Ampthill.

GENERAL SPECIFICATION

ENGINE

Cylinders	Four-in-line
Capacity	2393 cc
Bore/stroke	92/90 mm
Max power	112 bhp (84 kW) at 4200 rpm (DIN)
Max torque	183 lb ft (248 Nm) at 2400 rpm (DIN)
Block	cast iron
Heads	Aluminium alloy
Cooling	Water
Valve gear	Pushrod overhead valve, 2 valves per cylinder
Compression	22:1
Fuel system	Indirect fuel injection, rotary pump. KKK turbocharger, integral wastegate, air to air intercooler
Ignition	Compression ignition diesel
Bearings	Five main

TRANSMISSION

Drive	To all four wheels permanently.
Type	Five-speed synchromesh manual
Internal ratios and mph/1000 rpm	
Top	0.77:1/25.1
4th	1.0:1/19.3
3rd	1.39:1/13.9
2nd	2.13:1/9.1
1st	3.32:1/5.8
Rev	3.42:1
Transfer box:	1.22:1 (high) 3.32:1 (low)
Final drive	3.54:1

AERODYNAMICS

Cd	N/A

SUSPENSION

Front	Live axle located by radius arms and Panhard rod; coil springs
Rear	Live axle located by radius arms and A-frame, self levelling central strut; coil springs

STEERING

Type	Recirculating ball
Assistance	Yes

BRAKES

Front	Discs 30 cm dia
Rear	Discs 29 cm dia
Servo	Yes
Circuit	Front circuit duplicated
Rear valve	Yes

WHEELS/TYRES

Type	Aluminium alloy 6JK × 16 in dia
Tyres	205 16
Pressures F/R	
(normal)	25/35 psi 1.7/2.4 bar
(low load)	25/25 psi 1.7/1.7 bar

ELECTRICAL

Battery	12 v
Alternator	65 Amp
Fuses	5
Headlights	
type	Halogen
dip	110 W total
main	120 W total

GUARANTEE

Duration	12 months unlimited mileage, extendable to two or three years at extra cost
Rust warranty	Not specific

MAINTENANCE

Major service	12 months or 12,000 miles
Intermediate service	6 months or 6000 miles

COMPARISONS

Make/model	Price £	Eng. Cap. cc	Power bhp/ rpm	Torque lb/ft rpm	Max. speed mph	0-60 mph sec	30-50 mph in 4th sec	mph 1000 rpm	Over. mpg	Tour. mpg	Weight kg	Payload kg	Drag coef. Cd
Range Rover Turbo D	**18,109**	**2393**	**112/4200**	**183/2400**	**95.8**	**14.5**	**9.6**	**25.1**	**22.4**	**27.4**	**1938**	**2885**	**N/A**
Range Rover V8 petrol	16,886	3528	125/4000	190/2500	95.2	14.6	9.5	25.8	15.0	19.4	1860	2885	N/A

Make: *Land Rover* ***Model:*** *Range Rover Turbo D* ***Country of Origin:*** *Britain*
Maker: *Land Rover Ltd., Lode Lane, Solihull, West Midlands B92 8NW* ***Tel:*** *021-743 4242* ***Total Price:*** *£18,109.00*
Options: *Towing pack, £97; Electric windows, electric door mirror adjustment, power headlamp washers, door edge lights, £622; Alloy wheels, metallic paint, front mudflaps, £972.* ***Extras*** *fitted to test car: Electric windows, mirrors, door edge lights etc. Alloy wheels, metallic paint etc.*
Price *as tested: £19,703.*

BRIEF TEST

Range Rover High Line

Fuel injection, a new four speed auto — the world's best off-roader gets better.

Fifteen years on and the world's best four wheel drive is better than ever. So good, so capable is the Range Roger, that it's hard to believe the Japanese haven't yet tried to copy it.

There have been detail changes over the years, but the 1986 Range Rover is essentially the same car Land Rover Ltd presented to stunned motoring journalists back in 1970. A functional, yet stylish light alloy body rests on a box section steel chassis. Motive power is by way of the 3.5 litre V8 first used in the P6 Rover 3500 and drive is distributed permanently to both front and rear coil sprung live axles through a centre differential, which can be locked for extra traction in tough going. Brakes are disc all round.

Over the years have come a four door body, a five speed manual transmission, automatic transmission and improved quality control. With significant improvements to engine and interior and the introduction of a new automatic transmission, the 1986 version represents, by Range Rover standards, a major update.

The long serving 3.5 litre V8 is still under the bonnet, but Lucas-Bosch L-Jetronic fuel injection has boosted power by 22 percent. The engine now develops 112 kW at 4750 rpm and a useful 258 Nm of torque at 2500 rpm on unleaded petrol.

Five speed manual models now feature a new direct linkage, which has brought the gear lever closer to the driver and has improved the shift. The new automatic gearbox is ZF's impressive 4HP22 four speed, first seen in Australia in Volvo's 740 series and now adapted to the Range Rover's on/off-road requirements.

Engineering changes to the ZF 'box include a special torque converter housing and a new extension case which mates up to the Range Rover's LT230 two speed transfer case. A special sump and revised oil pick-up arrangement ensure a constant oil supply on steep gradients and side slopes, while all breathers have been mounted higher than usual to allow for deep water crossings. A gearbox oil cooler is also fitted.

A lock-up clutch by-passes the torque converter above 69 km/h to assist cruising and on-road fuel economy.

Other mechanical changes across the four model Range Rover line up, which starts at the $37,520 standard five speed version and stretches up to the subject of this test, the $45,770 Highline auto, have been restricted to detail suspension tuning. The rear coil springs are now dual rate and, combined with larger front radius arm bushes, are designed to increase roll stiffness. Damper settings have also been revised to suit the new springs.

Externally the '86 Range Rover, still one of the best looking off-roaders ever built, has been given minor cosmetic updates. The bumper end caps wrap around to the wheel arches, while a high-tech cathodic electro coat priming process has improved paint finish. Metallic colors feature a clear over base finish.

The contour of the panels covering the rear pillars has been changed to improve the fit with the rear body panels. Highline models feature a body protection moulding along either flank and a polyurethane front spoiler incorporating twin quartz halogen foglights, as well as alloy wheels.

Inside are more detail but significant improvements. The compact new ZF automatic gearbox has allowed Land Rover to develop a slimmer transmission tunnel for both automatic and manual versions, thus improving front seat legroom. The dash and fascia presentation has been cleaned up, with the radio now in the centre below the redesigned heater controls.

The interior treatment leaves you in no doubt that the Highline is the sort of Range Rover Princess Anne would use down on the farm. There are real tree wood cappings on the doors, as well as electric windows, air conditioning, central locking and electrically adjustable windows. Muddy gumboots would make a mess of the plush carpeting, though.

You sit high and mighty in comfortable velour trimmed bucket seats — complete with individual centre arm rests — and the Range Rover's huge glass area affords excellent visibility. The steering wheel is large but comfortable and the minor controls are better rationalised than in most British cars.

With power steering and the automatic transmission, the Highline breezes through city traffic. The fuel injected engine is torquey and responsive off idle, but acceleration is fairly leisurely. Left to its own devices the ZF gearbox tends to shift up early and the long travel accelerator makes it a little awkward to kick down. You won't leave any Brock Commodores — or even a well driven Gemini — standing at the lights, but keeping with the traffic flow is no problem.

On the highway, however, the Range Rover will cruise at 140 km/h with the poise and assurance of a European sedan. With a 0.7281 overdrive fourth driving through the 1.192 high ratio and 3.54 diffs front and rear, the V8 turns over at less than 3000 rpm at 100 km/h. Windnoise is remarkably subdued, given the Range Rover's square rigged bodywork, while the suspension is free of the rattle and clatter that accompanies most conventionally engineered Japanese off-roaders. The relative lack of wind, road and engine noise means the gear whine from the transfer case becomes intrusive about 60 km/h, though.

The suspension is not only quiet, it works superbly, too. Fifteen years on, the Range Rover's long travel four coil layout still sets the standard for four wheel drives. On the road it copes brilliantly with small bumps and large

PHOTOGRAPHY: DAVID G. SEGAL

"... and if you have to use first gear and low range, you don't, in the words of one JRA exec, really want to go there anyway ..."

holes, exhibiting a suppleness and control the Japanese have yet to perfect for their two wheel drive cars. Around town we found the ride impressive — and we'd just stepped out of a V12 Jaguar Sovereign. Say no more ...

On bitumen the Range Rover can be cornered quite briskly. There's plenty of lean, but the chassis, aided by the full time four wheel drive system, displays reassuring stability. The superb power steering allows the driver to accurately place the Range Rover in corners, while the well located rear axle refuses to be upset by mid-corner bumps. The tyres — Avon Rangemasters 215/75 R16 mud and snow pattern on our test car — define the Range Rover's cornering limits, provoking mild understeer at the limit.

In real off-road work the Range Rover is simply brilliant. Lock the centre diff and the damn thing will just about go anywhere — and if you have to use first gear and low range, you don't, in the worlds of one JRA exec, really want to go there anyway ...

The Range Rover makes off-roading easy — and comfortable. The supple, long travel suspension, which works so well on the road, helps the wheels follow the contours of the ground when the going gets tough. Traction is the key when it comes to serious off-roading — if the tyres fail to grip, either under acceleration or braking, even at 5 km/h, you'll probably get stuck.

The fuel injected engine is impressive off-road. There's good torque and the power delivery is nice and smooth. Off-

road purists will sneer at the Highline's self-shifting gearbox, but in truth the ZF makes it easier to walk the Range Rover over rough sections and to negotiate slippery surfaces without breaking traction on light throttle openings.

It may be almost foolproof, but there are some traps for young players. You must remember to lock the centre diff before you start into a tricky section, otherwise it only needs one wheel to break traction to bring the Range Rover to a wheel spinning stop. And it's best to use the auto 'box manually off-road. A lack of engine braking down steep slopes is traditionally a problem with automatic four wheel drives — Land Rover's engineers have got around it by modifying the ZF gearbox so it can be locked in first gear if necessary.

Few Range Rover Highline buyers will ever approach the limits of the vehicle off-road, however. Indeed, few Highlines will probably ever see a dirt road other

than on a run to the snowfields in winter. Like most prestige cars, Range Rovers are usually purchased for the image rather than the reality.

It's ironic, then, that one of the world's best off-roaders is most often seen loitering around fashionable shopping centres in Double Bay, but it's a fact importer JRA cheerfully accepts, even to the point of running ads extolling the virtues of buying second hand Range Rovers because ". . . many will have handled nothing tougher than the natural hazards of Toorak Road . . ."

Few companies are as keenly aware of the demands of their customers as JRA. Last year was the best ever for Range Rover in this country and JRA's car division director John Shingleton believes demand will continue to be strong in 1986, despite the slump in the overall market.

There's little doubt the Range Rover Highline will appeal to buyers looking for a prestige wagon type vehicle capable of carrying loads or towing a boat or horse float. That it's also a four wheel drive will only be seen as a bonus.

But the rising prices of the top line Japanese four wheel drives — Toyota's high roof Land Cruiser Sahara turbo-diesel currently weighs in at $39,498 — means the Range Rover is becoming a more realistic proposition for serious off-roaders. And the quality control has improved enormously since the bad old days of the mid-seventies. No longer is it the car only millionaires, Land Rover test drivers and motoring journalists can take off-road with disdain.

CAR DATA

MODEL	Range Rover High Line Automatic
COUNTRY OF ORIGIN	Britian
BODY TYPE	Wagon
SEATING CAPACITY	Five
PRICE	$44,220
OPTIONS FITTED	None
PRICE AS TESTED	$44,220
ENGINE:	
Location	Front
Cylinders	Eight
Capacity	3528 cc
Bore/stroke	88.9 x 71.1 mm
Block	Alloy
Head	Alloy
Valve actuation	OHV
Induction	Electronic fuel injection
Compression ratio	8.13:1
Power	112 kW at 4750 rpm
Torque	258 Nm at 2500 rpm
TRANSMISSION:	
Driving wheels	Four wheel drive
Gearbox type	ZF Four speed Automatic
Shift location	Floor
Gear ratios	
1st	2.4795:1
2nd	1.4795:1
3rd	1.0:1
4th	0.7281:1
Final drive ratio	3.54:1
BODY/CHASSIS:	
Construction	Welded box section
Material	Steel
Kerb weight	1971 kg
O/A length	4470 mm
O/A width	1780 mm
O/A height	1800 mm
Wheelbase	2540 mm
Front track	1470 mm
Rear track	1470 mm
Fuel tank capacity	81.5 litres
SUSPENSION:	
Front	Coils, with dual rate springs, radius arms and Panhard rod
Rear	Coils, with dual rate springs, trailing links
STEERING:	
Type	Recirculating ball, Power assisted
Turning circle	11.3 metres
Turns, lock to lock	3.25
BRAKES:	
Actuation	Hydraulic with servo assistance
Front	Disc
Rear	Disc
WHEELS:	
Material	Aluminium alloy
Diameter/width	75 x 16
TYRES:	
Make/type	Avon Rangemaster
Dimensions	215/75/16
PERFORMANCE:	
WEATHER	Cool
ROAD	Dry bitumen
ODOMETER READING	17,459
SPEEDOMETER ERROR AT:	
60 km/h	61
80 km/h	79
100 km/h	99
STANDING START:	
0-60 km/h	6.8 secs
0-80 km/h	10.9 secs
0-100 km/h	16.9 secs
400 metres	20.3 secs
Terminal speed	108 km/h
IN GEAR ACCELERATION:	3rd (or 'D' auto)
60-100 km/h	10.5 sec
80-120 km/h	15.6 sec
BRAKING FROM 100 km/h	
Average of four tests	46 metres

Melbourne specialist company Quadramotive has developed this Schuler Range Rover for a Middle Eastern customer. As Bob Maror reports, it isn't your ordinary 4x4 . . .

If Highline isn't high enough, you can always step up to a custom Schuler creation — such as this one which transforms your passe $46,000 Range Rover into a stretched four wheel drive limousine and reaches into the heady heights of the $100,000 bracket.

Built for His Highness the Crown Prince of Abu Dhabi, Shaik Kwaufa Bin Zaid Al Nghayan, this Range Rover is truly fit for a king. The extra twelve inches of wheelbase and body length are not there for an extra row of rearward facing seats. Instead, the length was added to enable His Royal Highness to launch his hunting falcons from a standing position in the rear, through the full length, electric power operated fabric sunroof.

But the length of this Hyper Rangie is merely its most visible attribute. Beneath the bonnet lurks a 355 cubic inch, stroked Brock motor built by HDT Special Vehicles and wielding some 300 horsepower and 400 ft.lb of torque — enough to chirp all four wheels from a rolling start on dry bitumen.

Running gear is equally impressive with a competition clutch feeding into a four speed transmission and lockable transfer case with Quaife torque sensing self-locking differentials front and rear. All of this, of course, is supported on uprated suspension using heavier rate coil springs and Bilstein gas shock absorbers.

Comfort for the royal posterior comes in the form of Recaro's full body adjustable electric power seats, including air adjustable lumbar support with all controls built into the thigh support padding. Elmo Swedish leather trim, Classic Image wool/nylon bonded carpet and Merino wool velour seat trims are set off by Quadramotive's full width walnut dash and glovebox.

With the extended length, the vehicle's ride is far superior even to that of the original Range Rover, and with so much muscle on tap, performance on the highway (and presumably in the sand deserts of Abu Dhabi) is astonishing.

Quadramotive Rangie has had an extra 300mm spliced into the wheelbase. Sunroof will allow new owner to launch his hunting falcons. Interior includes walnut dash, power Recaro seats.

AUTOMOBILE MAGAZINE

PREVIEW

RANGE ROVER

We try Europe's trendiest vehicle, as it sets sail for America.

BY DAVID E. DAVIS, JR.

British Air, Flight 0093, London–Detroit: There are rarefied social strata in Europe where arriving at some elegant gathering in a Range Rover would confer infinitely more status on one than arriving in something like a Countach or a Testarossa. I have just spent ten days in Great Britain putting over 1100 miles on a handsome green Range Rover, flogging around test tracks and wagon tracks and racetracks in a variety of Range Rovers and Land Rovers and sports cars, and spending almost every night in a different inn, club, or castle. Now I can relax in a large Super Club seat on my favorite airplane, the Boeing 747, and let the cheerful, competent people of British Air look after me for a few hours.

The rationale for this trip was based on the notion that J.L.K. Davis and I could bat around the United Kingdom and hobnob with people who do whatever people do with Range Rovers, to better understand the vehicle's function and philosophy. I can start by saying that the trip was a howling success. Based on our meetings and adventures I expect the Range Rover to take America's rich and famous—quail hunters, field trialers, fly fishermen, gentlefolk farmers, horsey people, and poseurs—like Sitting Bull took Custer. That is to say, emphatically and by surprise.

First Day: We arrive at Heathrow's spotless new Terminal Four, nine-thirtyish, sail through immigration and customs in about ten minutes, then drive our Range Rover into London, to the St. James Club, where we'll spend the night. We shower, change clothes, grab some lunch, and cab over to the Royal Geographical Society in Kensington.

We're greeted at the RGS by retired Brigadier Gathercole, the deputy director, and on our way up the stairs to his office I'm already overwhelmed by the place. After all, I'm the guy who has a quote from Sir Ernest Shackleton on his office wall, and here I was surrounded by the ghosts of Burton and Speke, Stanley and Livingston, Tenzing Norkay (Tiger of the Snows), and old Shackleton himself. The brigadier led us on a tour of libraries and map rooms and ramshackle offices that reminded one of nothing so much as the Circus, as portrayed in John Le Carré's books. I fully expected George Smiley to look up and smile vaguely as we passed through the map storage area in the cellar.

We were introduced to Ken and Julie Slavin, who act as expedition consultants for Land Rover/Range Rover, and we met a delightful young woman named Rebecca Ridley, who had only recently completed a four-month expedition to the Wahibah Sands of Oman as part of a Land Rover–mounted scientific team.

PHOTOGRAPHY BY COLIN CURWOOD

American pilgrim seeks insight among the Range Rover people.

Miss Ridley took us all back to the brigadier's office, where we had tea. On learning of my worshipful enthusiasm for Ernest Shackleton, she cried, "Oh, did you visit so-and-so's office on your tour? He has Shackleton's original plates." His false teeth? His china? I was enthralled. She dashed off and returned with a set of photographs from the 1914 Antarctic expedition that looked as though they might have come out of the lab the day before.

Second Day: Pack, eat breakfast, and roll downstairs with just enough time to cross St. James and wander up Jermyn Street for a block or two before our scheduled 9:30 departure. Bill Baker, our official U.S. Range Rover escort, goes into Turnbull and Asser and runs the old MasterCard right to the redline for some shirts and ties.

We depart London, find our way to the M1—Great Britain's first superhighway, completed in 1959—and set out northbound for Solihull, in the industrial Midlands, where Range Rovers and Land Rovers are built. Most motorway traffic bustles along at a smart eighty-five or ninety, in good-humored violation of the United Kingdom's seventy-mph speed limit. Our Range Rover sailed along, effortlessly running with the hot Fords and BMWs and big Mercedes that seem to set the pace all over Europe.

On the highway, the Range Rover behaves most like a tall luxurious hatchback sedan. Its full-time four-wheel drive goes unnoticed, but its four-wheel disc brakes are welcome indeed. It is quiet and smooth, and its ride has none of the choppy truckiness to which we Americans have become accustomed in our big Blazers and Broncos. One of its best features—for us, at least—was a relatively high sitting position combined with windshield and side glass that seem to come all the way down to the vehicle's (and the occupants') beltline. Visibility, as a result, is su-

perb, and the driver enjoys a sort of captain-of-the-ship feeling of being in command at all times.

Range Rover is produced by Land Rover U.K., Ltd., which is now operated as a separate company under the hegemony of the Rover Group, which used to be BL, which used to be British Leyland. Land Rover/ Range Rover no longer has anything to do with the manufacture of Rover cars, which are now themselves the product of a separate company, Austin Rover Cars.

After lunch in the executive dining room with Land Rover's affable senior management, we moved on to the shop floor where we watched Range Rovers being assembled and saw, for the first time, the American-spec Range Rover sitting off to one side in an inspection area. The differences between the European model and the U.S. model are not at first obvious.

The American-spec Range Rover corresponds, in the main, to a top-of-the-line five-door home-market model called the Vogue. It has a redesigned grille, the better to protect two fans and a heat exchanger and to generally increase airflow. It features a redesigned rear hatch that will operate more like the ones on American vehicles of this type, which is a good thing, because the hatch closure on home-market Range Rovers is a bit primitive. The instrument panel has been redesigned to accommodate revised air conditioning controls. Special Lytherm heat/sound insulation material is installed under all carpets to protect us Yanks from our catalytic converters and to make the U.S. model even more quiet than its parent. A removable American-style ball trailer hitch will be offered, inasmuch as the Range Rover is a very popular vehicle for towing boats and race cars and horse boxes in Europe. And, of course, the U.S.-spec vehicle will carry front, rear, and side marker lights to comply with our federal requirements.

The American Range Rover will also offer leather upholstery as a premium-priced option, but a four-speaker stereo system with rear-window-mounted antenna, a smaller (sixteen-inch) padded steering wheel with integral cruise control, electrically adjusted six-way power seats, and a sixty/forty split bench seat in the rear will all be standard equipment. The level of standard equipment is more like that of a luxury car than an all-wheel-drive sports-utility vehicle.

All U.S.-spec cars will be shipped with the ZF four-speed automatic gearbox and lockup torque converter. There will be no manual transmission offered, period. This breaks our hearts, because the five-speed manual offered overseas buyers is a good one, and offers a choice of ratios and a pleasing functionality that's exactly appropriate for the vehicle.

The automatic is probably every bit as good as the manual in off-road situations, but it just isn't as much fun, nor does it offer the same feeling of reassurance when one eases the front wheels over the edge of some precipice and looks straight down at the United Kingdom a quarter of a mile below. Range Rover defends its choice of the automatic on the grounds that too few Americans would order manuals to make the additional cost of U.S. certification cost-effective. If that's true, I am ashamed of my fellow colonials.

The Range Rover's 3.5-liter (215

cubic inch) V-8 aluminum engine is a peach. U.S. emissions control requirements have sapped fifteen of its horsepower, but it still produces a very useful 150 horsepower at 5000 rpm and 195 pounds-feet of torque at 2500. This engine was originally designed and produced by General Motors for the 1961 Buick Special, but the corporation was never able to make money on it and wound up selling it to Rover in 1967. It has been much refined and developed over the years. Rover used it in several cars, the Range Rover was more or less designed around it in the late Sixties, Triumph used it in the TR-8, Morgan uses it in the Plus Eight, TVR in its 350i and 390SE, and—most recently—it was installed in a new, fully modernized Land Rover which uses Range Rover suspension components as well.

In the details of its chassis and driveline, the U.S.-spec vehicle is virtually identical to the home product. This is as it should be, because the Range Rover, as sold since 1970, has built an enviable record as the world's most prestigious and successful all-purpose vehicle. Do not be put off by its luxurious personality. It is every bit as capable off-road—in the very worst kind of four-wheel-drive terrain—as it is on the boulevard. Only the Mercedes-Benz Geländewagen challenges it at this end of the market. And although the G-wagen offers the more sophisticated Steyr four-wheel-drive system with hydraulically controlled front and rear locking differentials, it lacks the Range Rover's ground clearance and suspension travel, and its delivery-van appearance may prevent it from ever catching the Range Rover in the European sales race.

Our factory tour and our inspection of the America-bound prototype completed, we went out into a light drizzle and were introduced to Mr. Don Green, whose title is Manager, Demonstrations, and who has taught—or attempted to teach—policemen, firemen, ambulance drivers, film stars, Third World potentates, members of the British royal family, and, yes, American journalists how to drive safely off-road. He was to take me onto the Land Rover Jungle and Land Track for a short review of the basics of all-wheel driving in difficult terrain.

I had previously driven Range Rovers on a couple of off-road adventures in Wales, and on a hunting trip in Austria, and felt quite at home as Mr. Green pointed me off into the haunts of coot and hern, with J.L.K. Davis and the estimable Bill Baker in the back seat.

Mr. Green lays down a few fundamental principles as we begin the drive:

"Don't wrap your thumbs around the steering wheel rim. A sharp impact on one front wheel or the other could render you incapable of ever again eating with chopsticks.

"Engage the differential lock whenever you head into terrain that could not be negotiated in a normal passenger car." (The diff lock works on the center differential, and can be engaged on the fly by simply pulling its short console-mounted lever toward the driver's knee. This lever also shifts fore and aft—a bit more complicated move requiring a well-timed double-clutch and a good ear—to select high or low range in the transfer case.)

"Select low range and the appropriate gear *before* entering a difficult section. Avoid the temptation to 'play safe' by selecting too low a gear when climbing.

"Low gear, low range, when descending all hills in difficult terrain, however steep, however long.

"Avoid the temptation to use either brakes or clutch when climbing or descending.

"Don't go fast. Films and commercials lead us to believe that we can drive across rough country at high speeds without penalty. This is not true. When you're halfway across the Sahara, on your own, and you break something . . .

"Always survey terrain about which you are uncertain on foot to reduce the risk of encountering an unexpected hazard."

I complete the course (watched by a handsome red fox who lives within the confines of the test track), to the accompaniment of Mr. Green's cool commentary. "That's it, sir, easy does it. Gently at the bottom now. Where there's water, there's a hole. Right sir, give it a bit of a kick up the pants. *With feeling*, sir!" And so it went. As the sun went down, I was given a diploma, the last line of

which said, "Never has so much been conquered." I wonder what it means.

Third Day: We drive to Eastnor Castle at Ledbury, in Herefordshire. Eastnor Castle is one of those architectural follies so dear to the hearts of rich, landed Englishmen of the nineteenth century. It was built during the Napoleonic Wars, in 1810, and is currently owned by Lady Somers, female descendant of the man who built it, and her husband, Major Ben Hervey-Bathurst.

The estate surrounding Eastnor Castle is part farmland, part hilly woodland, and it is here that Land Rover does a large part of its development and testing. This is also the place where Land Rover trains the crews for the Camel Challenge—in January, if you can believe it, living in tents and washing with cold water, in preparation for an event that usually takes place in some steaming jungle like Madagascar or New Guinea.

Mrs. Lindamood arrives, having been collected at Heathrow by photographer Colin Curwood, and soon loud noises and hearty laughter fill the castle courtyard. We spend the day with Don Green running J.L.K. Davis and Jean Lindamood through the same basic program that I had completed the day before, while I slog along in their wake at the wheel of a Land Rover One Ten and Colin Curwood takes pictures.

Secret of the Range Rover's off-road prowess is a stiff chassis and compliant suspension, with lots of wheel travel. The car simply walks over the rough stuff like a trials bike. Front suspension consists of coil springs, a Panhard rod, and massive forged radius arms that look to be about a yard long and as big around as Billie Jean King's lower leg. There's also a very welcome hydraulic steering damper.

Rear suspension consists of coil springs and lower links, with an ingeniously laid-out upper A-frame at the center of the axle, surrounding a Boge automatic self-leveling unit. Mounting all this hardware in the center of the rear axle affords extra rear wheel travel and allows the rear wheels to assume some unbelievable angles when crossing ditches and other obstacles. Range Rover proves in off-road utility vehicles—just as we learned in cars, many years ago—that coil springs are better than leaf springs.

Midafternoon, and we suit up to drive to Gatcombe Park, where we visit Captain Mark Phillips, husband of Princess Anne, world-class horseman, and enthusiastic Range Rover owner. The captain turns out to be an extremely pleasant person who loves his horses and the farm he runs, and who takes obvious delight in hauling us all over the property at flank speed in his blue Range Rover. Back at his house, he serves us tea and speaks fondly of his friend Jackie Stewart. Two very affectionate dogs, a lurcher and a corgi, join us for tea. I comment that I've always wanted to own a corgi, and Captain Phillips says, "You should speak to my mother-in-law. She has fleets of them." PR man Baker takes his picture, gets an autograph, and we depart.

Back at the castle by sunset, nobody remains but Mrs. Lindamood and the Davises, who dine with the major and his wife and sleep in sumptuous quarters in the family wing of the castle. The major is a great storyteller whose knowledge of cars and wars and the world is voluminous and whose wine cellar would bring tears to your eyes. When he's certain that we appreciate the stuff, he simply surrounds us with great and noble bottles, each one better than the one before.

At some point in the evening it turns out that for eleven months during World War II the major acted as a British military liaison with the French underground, working out of a village in the French Jura called Thoirette—the same village where J.L.K. Davis spent much of her time during the years she lived in France. They know the same bars, the same people, the same crossroads, the same bakers and storekeepers, and a great homecoming celebration erupts in French, while Lindamood, Lady Somers, and I shake our heads, grin, and drink more wine.

Fourth Day: There's something nice about waking up with the light streaming in through the windows of

your castle bedroom. Going down several flights of stairs to the kitchen where the lady of the castle is preparing breakfast—helped by two large dogs and the castle cat—isn't too shabby, either. As we eat breakfast, we watch a small fleet of Land Rovers and Range Rovers arrive outside, bearing our friends. The morning is a repeat of the previous day. More photographs, more howls of laughter, more slithering up and down the muddy hillsides in Don Green's Range Rover.

The afternoon is spent poking around the estate, visiting Ledbury to buy some Barbour gear, chatting with the major, and seeing Lindamood and Curwood off to London. Dinner that evening is at a nearby hotel, after which the major has arranged for me to go out shooting rabbits from the back of an open Land Rover with the older of his two sons, George, a respectable London businessman.

Large numbers of rabbits—considered pests in these parts—are dispatched, and we are careening across a meadow in pursuit of a fox when the Land Rover crashes into a hole nobody knew about. Son George smashes through the top bows and hammers the windshield frame with his upper mandible, while I go out over the tailgate feet first, with a shotgun in either hand. I am bruised and bleeding. George's lower face is a gory mess. The gamekeeper—our driver—seems to have sprained his thumb, and I, for one, wish he'd broken his neck. We go back to the castle, beaten men, where George is repaired by his mother and I am given whisky and have my wounds tended by my amused wife. The Land Rover will need some work.

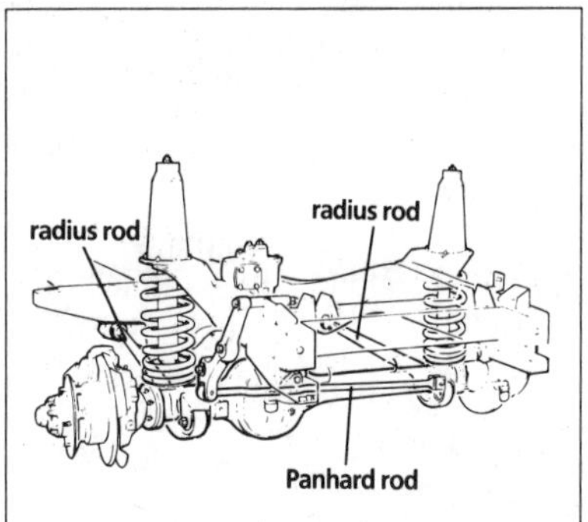

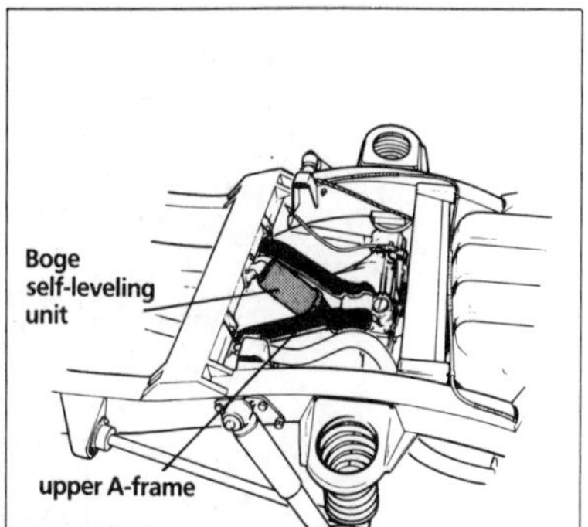

Sophisticated front and rear suspension (see text) are the basis of Range Rover's superior off- or on-road performance.

Fifth Day: Farewell to Eastnor Castle and the Hervey-Bathursts, whom we will miss. Ninety miles per hour for something over six hours from Ledbury to Perth, in eastern Scotland. We have a very nice picnic along the route, and get to toss the Range Rover through about fifty miles of winding, hilly two-lane road south of Edinburgh. We rent shotguns and buy ammunition in Perth, then drive north to Alyth and the Lands of Loyal Hotel, which turns out to be merely terrific.

Sixth Day: I am stiff but ambulatory. Paddy Featherstone-Godley collects us after breakfast and leads us out to Balintore Castle, a fine example of the nineteenth-century "Scottish Baronial" style, which is about to be divided into condominia. Balintore will be headquarters for a glorious day spent shooting—or shooting at—driven grouse and partridge. It's clear that there's been a slip-up. Paddy is driving a five-door Mercedes-Benz Geländewagen, which he adores, despite the fact that it looks like it might have been captured in the Falklands campaign.

Our friend Baker sleeps in, in preparation for some arduous shopping in far-off Edinburgh. He would meet us down the road that evening.

We are introduced to the other guns, a very cheery and friendly lot. Two of them are royal ministers, one being Tom King, secretary of state for Northern Ireland, which means that he rates two carloads of security people who can be seen guarding gates and blocking roads far down in the valleys below us as the day's shooting progresses. Paddy takes part of the party in his G-wagen, and we take the rest in our Range Rover. The hills are beautiful but wicked, and it is very civilized of Mr. Featherstone-Godley to have organized things in such a way that we can drive up and hike down.

I'm a little tense, acting as chauffeur for this group, which seems to a man to be enthusiastic Range Rover owners. Once when my timing is bad and I graunch badly going into low range, one of them says, "Mine makes exactly that same noise." We don't see many grouse, but America's honor is upheld when Mrs. Davis and I commit great execution among the partridges. We depart at day's end with two brace of grouse which we will present to Phil Llewellin, whose wife is one of Christendom's great cooks.

Seventh Day: We sleep at Sunlaws House, a beautifully done hotel, owned by the Duke of Roxburghe, who lives nearby in eighteenth-century Floors Castle. We arrived too late the night before for supper, but the hotel very kindly put a feast on a tray and left it, with a bottle of wine, in our room. Life is good. It is another day of brilliant sunshine, and we drive over to the castle to visit his grace, the duke.

At Floors Castle, the duke's

Trip Notes

Eastnor Castle, Ledbury, Herefordshire, England—A great castle tour most days in tourist season, with arms and armor, fine paintings, relics, tapestries, all the stuff you ever expected to encounter in a proper palace. There's a deer park, a camping ground, and miles of trails for hiking, bird watching, and encounters with members of Britain's elite SAS—who train in the area—and the equally elite engineers of Land Rover U.K., Ltd., who test here.

Sunlaws House Hotel, Kelso, Roxburghshire, Scotland—The Michelin describes Sunlaws as a Victorian country house. It is a former residence, age indeterminate, with comfortable rooms and excellent food. We stayed in a lovely twin-bedded room in a converted-stable annex and have every intention of returning. Sunlaws belongs to the Duke and Duchess of Roxburghe and sits in the valley of the Tweed, in the heart of the Scottish borders, reached by great enthusiast roads.

Floors Castle, Kelso, Roxburghshire, Scotland—Home of the Duke and Duchess of Roxburghe, Floors is some piece of real estate. It is a popular tourist attraction—more so since the word got out that Prince Andrew courted Fergie on the premises—and it was a setting for the film *Greystoke.* The original 1721 William Adam core must have been exquisite, but the century-later W.H. Playfair additions are a trifle Disneyesque.

Riber Hall, Matlock, Derbyshire, England—If you meet a beautiful woman of taste and wisdom in London, and she's willing to spend a weekend with you, take her to Riber Hall. Great food, a terrific wine list, and half-timbered rooms with canopy beds make this restored Elizabethan manor house a great destination.

Rowley's, Jermyn Street, London—Steak and French fries in the French tradition. You can choose a salad and pick a pastry from the cart, but you're there to sauté steak at your table, and you'll be glad you came. A block or so from Piccadilly, on the street that is the Intergalactic Capital of the men's haberdashery trade.

Lands of Loyal Hotel, Alyth, Perthshire, Scotland—Forty-five minutes north of Perth, an hour and a half from Edinburgh, Mr. Rennie, the landlord, dispenses a nice blend of hospitality and better-than-average one-liners. Surrounded by a network of winding roads that set American enthusiast drivers' mouths to watering, the Lands of Loyal is a former Scottish country house with nice, modern accommodations and first-rate country fare.

Range Rover is parked prominently outside the front door. We're ushered into a dim waiting room near the castle office, and Baker flips a light switch. Instantly, alarms go off all over the castle, and Baker is convinced that we're going to prison. The duke explains a few minutes later that it was merely the weekly test of the fire alarm system, but Baker has trouble regaining his composure.

The duke is a handsome young guy who serves us coffee and introduces us to a huge yellow Labrador retriever who sleeps in a basket by the ducal desk. We ask his grace about his life as the manager of what is both a huge agricultural property and a tourist magnet drawing thousands every summer. He makes it all seem pretty routine. We tour the castle, much of which was rather badly redecorated by his American grandmother, and then his gamekeeper, Jimmy Nairn, takes us for a tour of the estate in our Range Rover.

After lunch and some clay pigeon shooting at Sunlaws, we parted ways with Baker, and went south to England to meet Jean Lindamood, Kevin Smith, Georg Kacher, Mel Nichols, and Phil Llewellin. We drove through the Peak District as we went, and spent the night in Matlock at a delightful inn called Riber Hall, another Elizabethan house that just happened to have a first-rate restaurant.

The next two days were spent testing sports cars on the roads of the Peak District and at the Donington Park racing circuit, with the above-named, plus the great Derek Bell, perhaps the world's best sports car driver. That story will appear next month.

On the last day, returning to London in the middle of the morning rush on the M1, we were cruising as before in our green Range Rover—a steady ninety—when an unidentifiable silver car boomed up behind us and flashed its lights to pass. I whipped over to let it go by, and was passed by Kevin Smith and Georg Kacher in the new Rover Sterling, introduced on page 66 in this issue. Small world. We parked the Range Rover at Brown's Hotel in London, and really hated to give it up. It had gone everywhere with us, and done everything, and had really earned our respect. It is certain to be a smash hit when it arrives in North America.

RANGE ROVER

GENERAL:
Front-engine, 4-wheel-drive wagon
5-passenger, 4-door aluminum body
Base price (estimated) $30,000

ENGINE:
OHV V-8, aluminum block and heads
Bore x stroke 3.50 x 2.80 in (88.9 x 71.1mm)
Displacement 215 cu in (3528cc)
Compression ratio 8.1:1
Fuel system Lucas electronic injection
Power SAE net 150 bhp @ 5000 rpm
Torque SAE net 195 lb-ft @ 2500 rpm
Redline 5500 rpm

DRIVETRAIN:
4-speed automatic transmission
Gear ratios (I) 2.48 (II) 1.48 (III) 1.00 (IV) 0.73
Transfer-case ratios (High) 1.22 (Low) 3.22
Final-drive ratio 3.54:1

MEASUREMENTS:
Wheelbase 100.0 in
Length 175.0 in
Width 71.6 in
Height 70.8 in
Curb weight 4303 lb
Weight distribution front/rear 50/50%
Fuel capacity 21.0 gal

SUSPENSION:
Live-axle front, with radius arms, Panhard rod, coil springs
Live-axle rear, with triangulated upper link and trailing arms, coil springs, automatic ride leveling

STEERING:
Worm-and-roller, power-assisted

BRAKES:
11.7-in discs front
11.4-in discs rear

WHEELS and TIRES:
16 x 7.0-in cast-aluminum wheels
205R-16 Michelin X M + S tires

PERFORMANCE (manufacturer's data):
0–60 mph in 13.8 sec
Top speed 99 mph
Standing ¼-mile in 19.6 sec
EPA city driving (estimated) 13 mpg

RANGE ROVER OVER HERE

Roving with civility

BY JOHN LAMM

MAJOR HERVEY-BATHURST had just had us to tea and shown us around Eastnor Castle. One of the best collections of armor in England, they say. Ten minutes later, we were slogging through grimy mudholes in a Range Rover.

As we wound along a rutted road on a Welsh mountainside, the wind was trying to blow the rain through the windshield. The Range Rover ahead of us appeared to be half tipped over clambering up the rock-strewn trail; but when we went through the same maneuver, it didn't feel all that dramatic.

That's the way it seems to be with a Range Rover. A little island of civility in what could be a grimy world. And now these little islands are officially being exported to America.

Range Rover has always been something of a mystery in the U.S., so a little history is in order. Rover has been making automobiles since just after the turn of the century. Though its reputation for automobiles was dimmed here, most recently by the 3500 and earlier models too that had nothing to do with Land Rover Ltd, which was separated from the car division in 1978, Land Rover produces two vehicles, its namesake and the Range Rover. The former has been around since 1948, when it was introduced as a British replacement for the rugged Jeep. The latter was first sold in 1969, meant to be a more genteel Jeep . . . not a sissy, just more civilized.

Despite its immediate success overseas, the Range Rover wasn't sent to the States. The company had never really intended to export to the U.S., figuring volume would be too low to support the cost of certifying the vehicle and establishing a dealer network. Land Rover was also concerned that the Range Rover wouldn't fit in, considering the number of 4-wheel-drive utility vehicles already selling here at a price well below what theirs would cost. Owning an all-wheel-drive vehicle may be acceptable and even chic here now—note the Chevrolet Blazer shares a garage with a Bugat-

ti in ads—but it wasn't in the Seventies.

In the early Eighties, Land Rover reversed itself and began to develop the Range Rover for the States. Now the product seems to fit the market, an upscale utility vehicle at a time when Eddie Bauer and Banana Republic clothing stores can be found in the best shopping centers. What's more, being successful in America will be a bonus when state-owned Land Rover Ltd is privatized in the near future. For the U.S., a wholly owned subsidiary called Range Rover of North America was established to make certain a proper network of dealers and parts is in place before any vehicles are shipped from England.

Basis for a Range Rover is a sturdy-boxed, steel ladder frame. Mounted on it is a body—4-door versions only in the States—with a steel inner structure covered with aluminum body panels. The aluminum is for light weight and corrosion resistance.

Range Rover suspension consists of a live axle at each end, located by (respectively) leading and trailing arms, and with coil springs and tube shocks. There's also a Panhard rod in the front suspension. Range Rover engineers have combined relatively soft dual-rate springs and a great deal of wheel travel to create a vehicle that has both a reasonable ride on the road and the ability to work off-road without shaking your fillings out. One interesting element of the suspension is the large Boge Hydromat leveling device that is attached to the rear suspension through a sturdy A-arm at the front of the back axle. With the Hydromat keeping the Range Rover on an even keel, the pressure is off the springs, so to speak, to compensate for heavy loads, so the springs can be softer.

The full-time 4-wheel drive has high and low ranges through the center-mounted transfer case. It's possible to lock the case's differential if you need still more traction. Steering is by worm and roller, the brakes discs all around.

There were a number of changes made to the Range Rover to ready it for America. Naturally several things had to be done so it would meet U.S. regulations, such as side marker lights. We like our climatic comforts here, and the heater/air conditioner controls were redone. Other examples of changes include more heat/sound insulation and a removable-ball trailer hitch. We also like our music, so a better stereo system was ordered. A 6-way power seat replaces the manual version. A new steering wheel has buttons for cruise control, a feature that wasn't available before. And it was decided to offer leather upholstery. Being English, Land Rover uses Connolly hides, but this will be an option, the only option.

Naturally the V-8 needed certification, so Land Rover fits fuel injection—a Lucas L-Jetronic type—to replace the pair of carburetors used in the past. Compression ratio drops from 9.4:1 to 8.1:1 with a piston change. Catalytic converters are added to the exhaust system. Horsepower and torque don't suffer from the changes as much as one might expect, the former dipping from 165 bhp at 4750 to 150 at the same rpm, while torque slips from 207 lb-ft at 3200 to 195 at 3000. The small English firms, such as Morgan and TVR, have found these changes to the Rover V-8 quite a boon, because it makes using the engine in their U.S. models much simpler. The Range Rover's transmission is a 4-speed ZF automatic, which is all that will be offered here.

It's not unusual to hear the Range Rover called the Jaguar of utility vehicles. There is real wood on the dash, and with the leather upholstery the seats have the correct look and smell. However, the Range Rover is a utility vehicle, so you can't carry the Jaguar comparison too far. Naturally you sit up quite high in the 4x4. The seats are very comfortable, while the instrumentation is complete and easy to read. Controls have a light feel to them, but you can sense they are connected to rugged stuff, necessary for the sort of work Range Rovers are put through by such groups as the Royal Geographic Society and a number of the world's armies.

Most impressive of the Range Rover's virtues is its suspension, with its remarkable ability to get over rough trails with a minimum of fuss and jostling for the driver and passengers. And there's a certain bridge-like quality you can feel when the Range Rover clouts a rock or scraps its belly. No sissy dents. If there's any disappointment, it's with the power from the V-8, with a claimed 0–60 mph time of 13.8 seconds. One might expect a bit better, when considering the Range Rover's price.

This is when the other shoe falls. Expect the tab to run more than $30,000, but don't forget you can easily get Chevrolet/GMC Suburbans and Jeep Grand Wagoneers in the mid-$20,000 range. What's more, the Americans don't have a 3-year, 30,000-mile warranty.

If there's one shame in all this, it's that the very reasoning that finally brings the Range Rover to America—to fit in the lifestyle of the young affluent Rolex set—means the vehicles will probably be bought by people who will give them only light workouts. No running the Darien Gap. That's a shame, because it's like using a thoroughbred to give a kid his first riding lesson.

Range Rover

"Yo, Di. How 'bout we go shoot some quail?"

• "What is this thang, a Toyota?"

"No," we said, "it's a Range Rover. It's sort of like an air-conditioned dune buggy for people who drink bottled water and get expensive haircuts."

"You sure? It *looks* like a Toyota."

"Positive. You could buy three Toyotas for the price of this thing. Hell, you could buy a vacation home for the same money."

"What's it got makes it so special?"

"Full-time four-wheel drive, V-8 engine, lots of wheel travel, and plenty of snob appeal. When the queen of England wants to inspect the troops on field maneuvers, she usually shows up in one of these. Saudi princes use them for tailgate parties. That sort of thing."

"So?" he asks, looking us over pretty hard, implying that we don't look like either the queen or a Saudi prince.

"We're just driving it to keep them honest."

That scene, or something like it, took place at virtually every gas stop between Las Vegas and Ann Arbor. Most observers mistook the Range Rover for yet another Japanese import. That's not hard to understand, since the RR does look a bit like the offspring of an Isuzu Trooper and a Nissan Hardbody.

As new as the Range Rover may be to Americans, it's been part of Europe's stable of four-by-fours since its introduction in 1970. Over the years, the RR has been improved and refined with several updates: a four-door model in 1981, a five-speed manual transmission in 1983, fuel injection and a ZF four-speed automatic in 1985, and a host of minor changes. All along, the Range Rover was building itself an image as an off-road Rambo with the manners of an English country gentleman. It has become the vehicle of choice for the chap who doesn't want to get his Aston Martin or Rolls-Royce high-centered on a boulder while visiting the tenant farmers.

Before leaving for Michigan, we got a chance to play with our Range Rover at the bottom of the Grand Canyon. There we found that a lot of the Bigfoot-in-a-tux hyperbole that one hears about this vehicle is well deserved. We crawled up dunes, forded the Colorado River, bumped over boulders the size of large doghouses, and blasted down miles of gravel roads, and the Range Rover never once bogged down, got tippy, or rattled our teeth loose. Even at 70 mph over unpaved roads, it had a remarkably smooth ride, more like a Cadillac than a dirt donk.

Part of the RR's poise over the rough stuff is due to the huge coil springs positioned between its rigid axles and its ladder frame. The front spring rates are linear, but the rear springs are dual-rate, starting at 123 pounds per inch and rising to 170. These long and compliant coils provide eight inches of vertical wheel travel at the front and a whopping eleven inches at the rear. The resulting suspension suppleness allows the Range Rover's wheels to conform like the tracks of a tank to the most tortured terrain without bouncing the passengers off the cabin walls, spilling their cocktails, or causing them to blow their cool.

A further refinement is the use of a Boge Hydromat ride-leveling unit in the rear suspension. This device looks like an extralarge shock absorber but performs a somewhat different function. Mounted between the frame and the axle housing on an A-arm (which also serves a locating function), the Hydromat unit uses pressurized gas, hydraulic fluid, and the pumping action of normal suspension travel to help the rear springs support the body on an even keel, no matter what load is being carried. The Hydromat automatically senses a low ride height and closes an internal valve to trap hydraulic fluid above a piston in such a way that a leveling force is produced. If a load is removed and the ride height is sensed to be too high, the valve opens, hydraulic fluid passes through the piston, and the leveling force is reduced.

The Range Rover's full-time all-wheel-drive arrangement isn't exactly cutting-edge technology, but it gets the job done, both on- and off-road. The ZF four-speed automatic (the only transmission available)

drives a transfer case that delivers power to the front and rear axles by means of a center differential and two driveshafts. The transfer case also has a high and a low range and a means of locking the center differential. In high range, the diff lock can be engaged at any speed by flicking the transfer-case lever to the left. To engage the low range, the Range Rover's speed has to be no more than about 5 mph. You shift the transmission from drive to neutral, push the transfer lever from high to low, and then shift back into drive. In the Superman mode, you can leap tall mountains or flatten your tenants' sod hovels when they start getting uppity. A warning light on the dash illuminates whenever you lock the center differential.

Despite its extensive use of aluminum body panels, the Range Rover is a 4308-pound chubbette. A major contributor to this obesity is the massive boxed-section frame; with the body off, it looks like something a civil engineer would design to span a river. With so much weight to haul around, the slavish devotion of every one of the 150 horses produced by the 3.5-liter aluminum V-8 engine is required to keep the RR rovering along. There is ample torque on hand, but the RR needs a long runway to achieve takeoff velocity. It requires 13.2 seconds to reach 60 mph. The quarter-mile eats up 19.2 seconds, the Range Rover rolling through the traps at a leisurely 72 mph. In our testing, we found a terminal velocity of 95 mph in third gear. The ZF transmission has a

kickdown switch located under the throttle pedal, so you have to lift slightly for an upshift to fourth gear.

Once up to speed, the Range Rover is a comfortable, if not particularly aggressive, highway cruiser. Without a whole lot of reserve power on hand, you have to plan your passes carefully. You also have to think ahead if you want to conquer steep grades in a single bound. The drill is to build up a lot of momentum where the road is still flat; then you carry as much speed as possible up the hill. If you rely on the standard-equipment cruise control, you'll find yourself consistently losing five to seven miles per hour when climbing a long or steep grade. If you lose too much speed, the transmission eventually kicks down for the surge to the top. Short stops also demand planning. It takes the four-wheel disc brakes a lengthy 223 feet to stop the RR from 70 mph.

The Range Rover's cockpit is comfortable, though not terribly spacious. The front seats have power adjustments for height, seatback tilt, and fore-and-aft position, and they are equipped with adjustable armrests, one per chair. The rear-seat passengers get two armrests per seat. Wherever you sit in the Range Rover, you get a mightier-than-thou view of the world. You're way up there off the pavement, Commander Cool trading steely looks with the lowlifes in pickups.

Unlike some of the recent Japanese entries into this market, the Range Rover offers few spaces in which to stuff the things you tend to accumulate on long road trips, or even in daily driving. Other than a bin in the center console and a shallow map pocket in each door, there is no place to stash things for easy retrieval on the move. Road drinks invariably end up between your legs, giving you a terminal case of inner-thigh freeze.

While generally comfortable, the RR's interior is not a resounding ergonomic success. The radio, for instance, is hard to reach and requires taking your eyes off the road for too long a time. The horn button is located at the end of the turn-signal stalk, hardly the ideal location.

A few other areas could also use improvement. The plastic doesn't have the quality you expect in a $30,000 machine. The unvarnished wood trim pieces along the tops of the doors look like shop-class rejects. Some people may find the wood quaintly evocative, but to us it looks like an afterthought.

The Range Rover's suggested retail price is $30,825. For that you get a long list of standard equipment, including

Technical Highlights

• The introduction of the Range Rover to these shores brings home an engine born in the glory days of the American car industry. The 3.5-liter V-8 that powers the RR was designed by Buick engineers and introduced in 1961. Powering compact sedans from Buick, Oldsmobile, and Pontiac, it was the world's first all-aluminum automobile engine produced in high volume. Moreover, in the 1962 Olds Jetfire, with modifications and water injection, it was also the first production turbocharged engine.

At 350 pounds, the Buick V-8 was relatively lightweight, but it was expensive to produce at a time when customers and the industry were at least a decade away from discovering the benefits of lighter automobiles. As a result, it was retired from service in 1964, replaced by cheaper engines that were heavier but simpler to manufacture. Two of these evolved directly from the aluminum V-8: Buick's 90-degree V-6, which was introduced in 1962 in 3.2-liter form and is the ancestor of all current Buick V-6s, and a 4.9-liter V-8, which was produced from 1964 to 1979.

Although the alloy V-8 was too sophisticated for Detroit, racers all over the world found it much to their liking. Mickey Thompson hopped one up for the Harvey Aluminum Special, which Dan Gurney drove in the 1962 Indy 500. Lance Reventlow used versions of the engine in his Formula 1 cars, and Bruce McLaren fitted one to the Zerex Special, arguably the first Can-Am car. The engine even found its way to the top of Formula 1: the block of the Oldsmobile version was the foundation of the Australian Repco engine that powered Brabham cars to World Championships for Jack Brabham and Denis Hulme in 1966 and 1967.

The merits of the aluminum V-8 were noticed by more than just racers. In 1965, when British Leyland needed a new engine for its Rover line of automobiles, it purchased the rights to manufacture the Buick engine from GM. Since GM was using most of the original tooling to produce iron V-6 and V-8 engines, Rover was free to modify the engine to suit its manufacturing facilities and lower production needs. The two most important changes were the switch from cast-in to pressed-in iron cylinder liners and the adoption of sand casting instead of permanent-mold casting for the engine block.

In this Anglicized form, the aluminum V-8 has powered top-of-the-line Rovers from the 3500 sedan of 1969 to the Range Rover of today. It has also been used in the MGB, the Triumph TR8, and the odd Morgan and TVR. Success in everything from sports cars to off-road vehicles, with a pair of Formula 1 championships thrown in, is not bad for a powerplant that GM gave up on 23 years ago. *—Csaba Csere*

power seats, power windows, air conditioning, a Clarion four-speaker AM/FM-stereo/cassette unit, cruise control, power rear-view mirrors, a rear wiper-washer, and a towing package. The only option is leather upholstery. If your priorities are snob appeal and playing one-upmanship down at the polo club, the price should be an attribute, not an obstacle. If, on the other hand, you're looking for genuine value, efficient packaging, and modern mechanicals, there is no shortage of alternatives in the dual-purpose market. If you really have to have archaic mechanicals and something with a lot of heft, you could slide yourself into a comparably equipped Jeep Grand Wagoneer for about $5000 less. But it's tough to pretend you're Lord Duke Slapsaddle when you're driving something built in Toledo.

—Tony Assenza

COUNTERPOINT

• I'd built up quite an image of the Range Rover long before I ever had a chance to drive one. I used to watch John Steed of "The Avengers" catch crooks with one. I'd seen films of Britain's SAS commandos performing exacting counterterrorist maneuvers with Range Rovers by their sides. And then there was the reputation for how luxurious this workhorse was. I just knew the Range Rover had to be the quintessential four-wheel-drive machine.

My expectations are only partly fulfilled. The Range Rover is no more luxurious than any number of big four-by-fours. Irritating mechanical noises intrude from the engine compartment. And the thing is awfully expensive. I think $35,000 is a lot to ask for a machine that is, at best, only marginally better than its competition.

I don't think you'd be crazy to spend the bucks if your heart is set on this veddy British four-wheeler. But now I know that John Steed didn't always get the bad guys just because he had the better wheels. *—Arthur St. Antoine*

Parnelli Jones once said that racing down Baja is "like being in an all-day plane crash." For off-roaders sorely tested by tough terrain transmitting itself into their, uh, cabs, the Range Rover is the only answer. This thing rides better than Preparation H.

Off-road, with viscous shock damping and long, wonderfully elastic spring travel, the Rover leaves the Neanderthal four-by-fours wincing in its wake. The only thing that feels like the onset of an all-day plane crash is climbing in and swacking your knees on the low steering column. For everyday driving, the superb chassis dynamics, the fine steering, the true tracking, and the tall glass make you feel as if you're riding the rails in an observation car. Normally can't keep tabs on shuffling traffic? No worries, mate: the Rover seats you just high enough for a comfortably layered view.

Too bad the little motor feels like an eggbeater with too many eggs in its basket. With an all-aluminum 350 Chevy up front, more than the ride would take things in stride. Then Rover could claim, "No pain, all gain."*—Larry Griffin*

I can't get too worked up over the Range Rover. When it was introduced in 1970, it immediately became the definitive blend of off-road capability, on-road comfort, and high style. In the intervening decade and a half, though, most other off-road machines have polished up their manners considerably.

The Range Rover's coachwork no longer looks particularly futuristic or sleek, its driveline lacks modern limited-slip technology, and its luxury features and power assists have been matched by any number of current off-roaders. And the Range Rover is no longer the only four-wheel-drive machine with a decent highway ride. Indeed, now that countless four-wheel-drive cars are available, there is no reason to drive a machine like the Range Rover just to have secure traction on slippery roads.

Perhaps I would feel differently if I trekked into the bushes every weekend to make like an antiaircraft battery whenever any ornithological creature flew overhead. For the kind of use most four-by-four owners have in mind, though, there are many better choices.

—Csaba Csere

ehicle type: front-engine, four-wheel-drive, 5-passenger, 5-door wagon

rice as tested: $31,375

ptions on test car: base Range Rover, $30,825; freight, 550

tandard accessories: power steering, windows, seats, and ocks, A/C, cruise control, rear defroster and wiper

ound system: Range Rover/Clarion NT2 AM/FM-stereo adio/cassette, 4 speakers

NGINE

ype .... V-8, aluminum block and heads
ore x stroke .... 3.50 x 2.80 in, 88.9 x 71.1mm
isplacement .... 216 cu in, 3532cc
ompression ratio 8.1:1
uel system .... Lucas electronic port fuel injection
missions controls .... 3-way catalytic converter, feedback fuel-air-ratio control, EGR
alve gear .... pushrods, hydraulic lifters
ower (SAE net) .... 150 bhp @ 4750 rpm
orque (SAE net) .... 195 lb-ft @ 3000 rpm

RIVETRAIN

ransmission .... 4-speed automatic with lockup torque converter
ransfer-case ratios, high range 1.22:1
low range 3.32:1
inal-drive ratio 3.54:1

ear	Ratio	Mph/1000 rpm	Max. test speed
L	8.23	2.9	16 mph (5500 rpm)
L	4.91	4.8	26 mph (5500 rpm)
I L	3.32	7.1	39 mph (5500 rpm)
V L	2.41	9.8	54 mph (5500 rpm)
H	3.02	7.8	43 mph (5500 rpm)
H	1.81	13.0	72 mph (5500 rpm)
I H	1.21	19.5	95 mph (4850 rpm)
V H	0.89	26.5	94 mph (3550 rpm)

IMENSIONS AND CAPACITIES

Vheelbase .... 100.0 in
rack, F/R .... 58.5/58.5 in
ength .... 175.0 in
Width .... 71.6 in
Height .... 70.8 in
Frontal area .... 29.1 sq ft
Ground clearance .... 7.5 in
Curb weight .... 4308 lb
Weight distribution, F/R 48.7/51.3%
Fuel capacity .... 20.0 gal
Oil capacity .... 6.0 qt
Water capacity .... 12.0 qt

CHASSIS/BODY

Type .... full-length frame with rubber-isolated body
Body material .... welded steel and aluminum stampings

INTERIOR

SAE volume, front seat .... 54 cu ft
rear seat .... 39 cu ft
trunk space .... 16 cu ft
Front seats .... bucket
Seat adjustments .... fore and aft, seatback angle, front height, rear height
General comfort .... poor fair **good** excellent
Fore-and-aft support .... poor fair **good** excellent
Lateral support .... poor **fair** good excellent

SUSPENSION

F: .... rigid axle, 2 leading arms, Panhard rod, coil springs
R: .... rigid axle, 2 trailing links, 1 central A-arm, coil springs, 1 self-leveling device

STEERING

Type .... recirculating ball, power-assisted
Turns lock-to-lock 3.4
Turning circle curb-to-curb .... 39.0 ft

BRAKES

F: .... 11.8 x 0.6-in disc
R: .... 11.4 x 0.5-in disc
Power assist .... vacuum

WHEELS AND TIRES

Wheel size .... 7.0 x 16 in
Wheel type .... cast aluminum
Tires .... Michelin X M+S 200, 205SR-16
Test inflation pressures, F/R .... 28/38 psi

CAR AND DRIVER TEST RESULTS

ACCELERATION Seconds

Zero to 30 mph 4.2
40 mph 6.4
50 mph 9.3
60 mph 13.2
70 mph 18.4
80 mph 26.8
90 mph 41.0
Top-gear passing time, 30–50 mph 6.9
50–70 mph 10.9
Standing ¼-mile .... 19.2 sec @ 72 mph
Top speed .... 95 mph

BRAKING

70–0 mph @ impending lockup .... 223 ft
Modulation .... poor **fair** good excellent
Fade .... none **moderate** heavy
Front-rear balance .... poor **fair** good

HANDLING

Roadholding, 300-ft-dia skidpad .... 0.66 g
Understeer .... minimal **moderate** excessive

COAST-DOWN MEASUREMENTS

Road horsepower @ 30 mph .... 9 hp
50 mph .... 24 hp
70 mph .... 50 hp

FUEL ECONOMY

EPA city driving .... **13 mpg**
EPA highway driving .... 15 mpg
C/D observed fuel economy .... **17 mpg**

INTERIOR SOUND LEVEL

Idle .... 50 dBA
Full-throttle acceleration .... 74 dBA
70-mph cruising .... 71 dBA
70-mph coasting .... 70 dBA

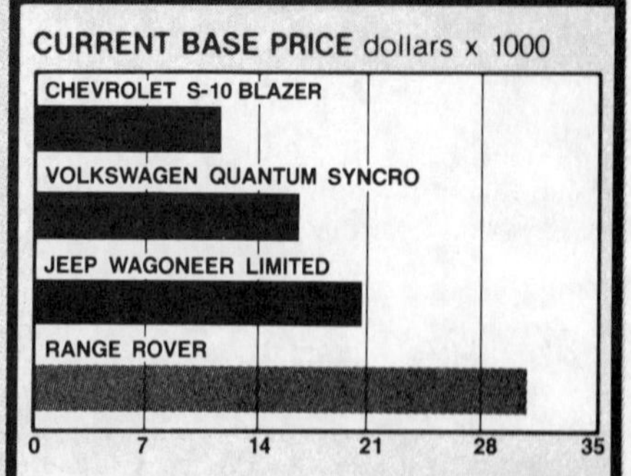

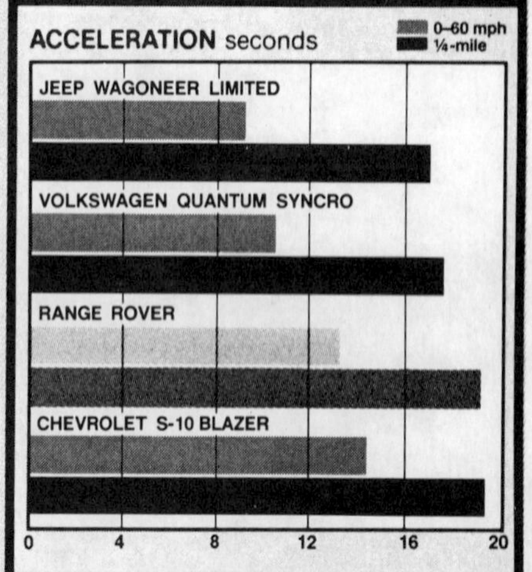

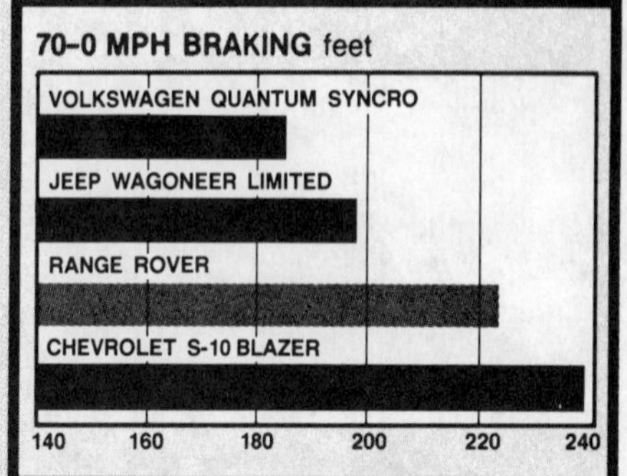

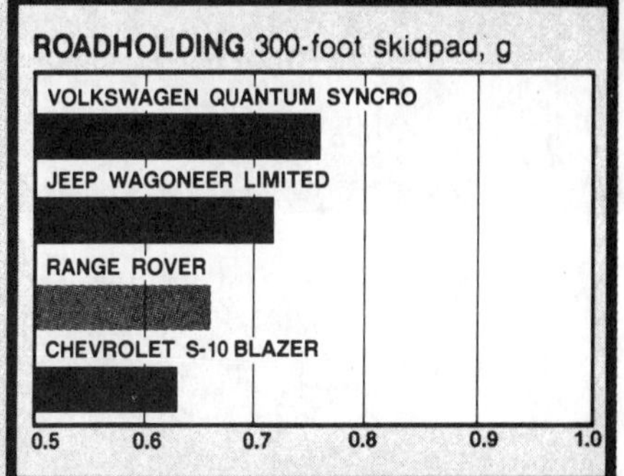

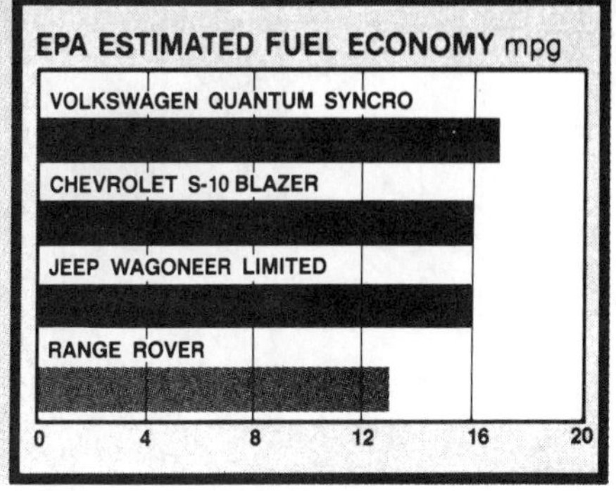

Carter's highland fling in his loaner Range Rover took him through all kinds of Australian Alpine terrain and was fun to drive, scenic, and a real work-out for the vehicle

HIGHLINE IN THE HIGHLANDS

Driving the 1987 Range Rover Highline is like travelling first class on an international Jumbo jet, reports Jeff Carter. His memory lane takes him high into the Victorian mountains

YEARS AGO I was lucky enough to have a two year test drive of one of the original pair of Press Range Rovers imported by what was then the British Motor Corporation. That was about 15 years ago. It was red, and had been test driven by a host of Australian magazine and newspaper writers when I got it. Despite the drubbing it had suffered "Big Red" was in faultless condition and continued thus for the hundred thousand or so kilometres I drove it all over Australia.

What am I doing, talking about four-wheel drive vehicles, when I swore off them back in 1976? Well, a four-wheel drive seemed the logical choice for my fifth journey down Memory Lane — a safari to Victoria's wildest mountain country, where my father — Pop, we called him — used to take me on trout fishing and deer hunting trips during World War Two. A 1987 Range Rover Highline seemed a most appropriate selection for this adventure.

Wartime petrol rationing made it difficult to get any distance away from Melbourne, but Pop's factory was supplying equipment to the Yanks and they had plenty of petrol. Plenty of everything in fact: grog, ammunition, cigarettes, silk stockings . . . and most amazing vehicles, called Jeeps. These were something entirely new on the local scene and from our cargo-cult viewpoint, their ability to climb around in the toughest country was miraculous. When the first Chev blitz wagons arrived, we gaped like neanderthal man confronted with Concorde.

Pop was delighted with the Yanks, because they could take him where he wanted to go — and the Yanks were delighted with Pop, because they needed a guide to take them into our huntin' an' fishin' country. Pop had done it all before, the hard way, in late '20s and early '30s Chevs, Fords, Whippets and "bull-nose" Morrises. Now he was in his element as expert local guide, in charge of a convoy of Jeeps and sometimes a Chev blitz wagon.

One of his favourite venues was centred on what is now Eildon Weir. We used to go via Healesville, Buxton and Taggerty, cross the Rubicon river to Thornton and then make our way by various bush tracks to fish the Delatite, Big River, the Howqua or the Goulburn near Jamieson.

It was big, steep, virgin country that tested even the Jeeps. After storms, fallen trees blocked the tracks and had to be chopped out of the way. To my young eyes, it was wilderness without parallel: rushing streams below soaring mountains clad with tall eucalypt forests; small clearings here and there, usually around a tiny split-log cabin or corrugated iron hut with smoke rising from the frogmouth fireplace.

Our chief quarry were rainbow and brown trout, which Pop and his mates pursued relentlessly with spinners, flies or bait, depending on the season.

Occasionally we stalked deer in the forests on the upper Howqua and

The scene above is on the way to the gold-mining settlement of Happy Valley, in country so wild that only the lure of gold opened it up. At left, this great stand of trees is in the snow country on the road above Dargo, heading towards Mount Hotham

◄ Jamieson rivers, but without dogs these forays were mostly unproductive. I was glad, because although I enjoyed the excitement of sneaking through the trees and undergrowth, I did not relish the spectacle of large animals dying.

It was high adventure for young Carter, but after I had seen one or two deer killed, I was always on the side of the hunted, not the hunters.

I set off from Sydney for Victoria's high country one Friday in June. As I sped effortlessly and luxuriously down the Hume Highway toward Victoria it was immediately obvious the 1987 Range Rover Highline was a much nicer conveyance than "Big Red". Comfort-wise, on bitumen — I had as yet no experience of my new steed in off-road conditions.

It was rather like travelling first class on an international Jumbo jet, surrounded on all sides by luxury: warm air, soft music, subdued lighting, lots of buttons and knobs to press, pull or slide, all experienced with little sense of motion. I half expected a steward to arrive with a tray of drinks!

The velour-lined seats, adjustable every which way, are truly luxurious, with drop-down central arm rests (also adjustable) and a full tilt-chair mechanism that is difficult to describe but delightful to experience. There is deep pile carpet under-foot, easily removed to reveal rubber matting for muddy days. The door sills are walnut capped with electric windows (the fastest I've encountered) and wonder of wonders: when you open the doors, "puddle

Carter's country may be rugged, but it is not entirely unpopulated. This picture is of fellow campers near Howqua, where a gourmet's retreat offers its select clientele a somewhat less spartan alternative lifestyle

lamps" concealed in the lower sill come on, so you can avoid stepping into water, mud, or a lurking cowpat!

All windows, fore and aft, can be controlled by driver or front passenger by buttons on the transmission hump console. Rear passengers have their own window controls, but if they are small children, the driver or co-pilot can press a button on the console which prevents the rear windows opening more than half way. Both external rear vision windows are electrically controlled and heated.

There is no fascia glovebox, but a deep, lidded box on the transmission hump behind the hand-brake serves the same purpose. Air-conditioning is integral in the Highline and is excellent. So is the heating and demisting system, once mastered. Likewise the four-speaker cassette-radio, complete with station-seeking and seven station memory in both AM and FM modes. You look in vain for a radio aerial, which is built into the rear window. Fingertip stalk controls for lights, indicators, wipers, washers, etc. are functionally well placed. Instrumentation is excellent, with *fifteen* warning lights in addition to the tacho, speedo, temperature and fuel gauges.

The aluminium fuel-injected V8 motor sounds beautiful and I'm sure will prove as durable and reliable as the original version. It produces 110kW (150bhp) if that means anything to you. More to the point, it pushes the RR to 60 km/h in first gear, 90 km/h in second, and 140 km/h in third — with the tacho at 5500rpm in each case and not anxious to go further.

In fourth, with the tacho a smidgin past 5000rpm, you get an indicated 170 km/h. In fifth, at a bit under 5000rpm, you can also have 170 km/h rather more quietly. I learned these interesting facts on Friday evening when the Yass district constabulary was presumably having its dinner break.

After a Vietnamese meal in Yass I reached Albury in time to learn from the late television news that another arrival was bad weather, in the form of rain. Further south, where I was bound, it was snowing. At Benalla next morning, my luck was out. Drizzling rain from leaden skies, with cheerful radio news readers telling me of sleet and snow along the Dividing Range from Eildon to the Dargo high plains.

Lacking the imagination to change my plans, I pressed on. To Mansfield, via the Midland highway, then more excellent bitumen through steadily rougher country to the delightful mountain town of Jamieson, crossing the Delatite and Howqua rivers on the way. The RR fairly sailed along through the steady drizzle, happy as Larry, devouring the increasingly twisty wet bitumen sure-footedly. Yes, the cabin still seems to lean gently when cornering, but apprehension gives way to confidence when it becomes obvious the four driven wheels aren't going to budge off line until you do something really adventurous.

At the Howqua bridge crossing, a gravel road beckoned me upstream. After half a dozen kilometres, it began to deteriorate and twist and wind through the trees, following the course of the river. Ah, here was a taste of the old days. Then, lo and behold, a fine homestead behind an imposing gate with a sign announcing "Howqua Dale Gourmet Retreat — entry by prior booking only". Not at all like the old days!

Up-river another dozen kilometres, over a tortuous, single-lane track, very wet and muddy, to Fry's crossing where I had several times camped and fished with Pop. It was a comparatively slow but effortless trip for the RR and despite several deep gutters and mud-bogs, a cup of tea sitting in its saucer would have survived on the lid of the transmission hump glove box. There were three groups of campers near Fry's, all making the best of the wet conditions around large fires, assisted by warming beverages, mostly bottled.

At Jamieson, darkness came and I was obliged to camp ignominiously in a public park a few kilometres out of town toward Kevington. It was raining heavily and I was glad of a shelter shed where I could brood on my misfortunes. While thus engaged, I recalled sighting a cosy pub back in Jamieson — and had there not been a sign in a side street saying "Restaurant"? Always the investigative journalist, I drove back.

The pub was quite large but made cosy by a huge roaring log fire beneath a mantle bedecked with dog traps and various other items of machinery and memorabilia associated with hunting, fishing, gold mining and grazing. Stuffed deer heads stared glassily down from the walls at the drinkers, diners and pool players.

Eventually I stumbled out into the sheeting rain and splashed my way a hundred metres to the restaurant located in an English cottage-style building under some big oaks. My luck had changed, I had discovered a proper restaurant, "Watson's of Jamieson", presided over by engaging, inimitable Milton Watson. If you're ever in Jamieson, don't miss a meal at Watson's. The building is interesting, a replica of the "Shere Inn" in England, built by one Alan Macgillycuddy Figgis from mud bricks made on the site and stone from the nearby "Star of the West" gold mine at Kevington.

On Sunday, it took the whole day to drive little more than one hundred kilometres to my next camp, not far past Walhalla, an historic mining centre. I did stop to take a few photos and once I had to chainsaw a fallen tree out of my way. But this is one helluva twisty road and what with rain and fog, there was only one way to travel — very slowly! It was child's play for the RR, no test at all, but with blind hairpins at about one-a-minute, I wasn't going to test its cornering abilities!

What the day did prove was the excellence of the RR's power steering, which gives plenty of road feel, despite being light. The journey was tedious in the extreme, but effortless physically. What possessed men to make tracks into outlandish country like this? The answer is gold. Just about every settlement from Jamieson to Walhalla and north-east through Dargo to Omeo and beyond was pioneered by miners. And there's still gold in them thar hills, some men claim . . .

It was a great road, particularly the first section to Gaffney's creek where the road parallels the Goulburn river and provides easy access to countless camping sites in small clearings among the tall eucalypts. No facilities, of course (who needs 'em!) but plenty of firewood and excellent trout fishing within micturating distance.

I camped in a deep gully and spent the night listening to the wind rage like an express train through the trees on the ridges far above me — and the steady patter of rain on my tent. Next day at Maffra, a sizeable town which had suffered extensive storm damage during the night, I replenished my supplies and headed for Dargo via Briagolong and Culloden on Freestone creek. Another helluva road, as tortuous as the creek it follows. Picturesque, but so heavily forested there are few possibilities for camping. At Cobbannah I joined the mostly bitumen road linking Bairnsdale and Dargo. The country opened out considerably and between rain squalls I glimpsed magnificent vistas of broad cleared valleys below forest clad ranges. Also a few brace of emus.

Dargo is still basically a store, pub and sawmill. Serious drinkers should play it safe and BYO, because the pub is inclined to be shut when you need it most. By contrast, the store is just what you would expect of a high-country outpost, complete with genial, helpful proprietors — plus a slow-combustion stove, very welcome on a cold, wet June afternoon. Armed with a detailed map prepared by the Victorian Mountain Tramping club and bought at the store, I set out for the pioneer gold mining field at Happy Valley, in the Wonnangatta-Moroka National Park.

This had been the scene of some of Pop's deer hunting, long before the creation of the park. The gravel road follows the course of the Wonnangatta river and is called Crooked River Road for obvious reasons.

Late in the afternoon I took the Mount Grant fork and drove about six tortuous kilometres to Randall's homestead. A bit further on I was confronted by a deep ford. The light was failing, but as far as I could make out, the crossing would be a very good test indeed for the Range Rover. Some startled cattle plunged into the water and tried to get across, then thought better of it and thrashed away downstream. I peeled of my boots and jeans and waded in until the most sensitive portion of my anatomy recoiled from the icy touch of the swirling dark water. Dumb animals often show good sense, I mused — and retreated to dry land like the cows.

Back on the Happy Valley road, I crossed Kingwell's bridge and went on to the National Park and a good camp on the banks of the Wonnangatta. It was almost dark and raining, but with the aid of my trusty chainsaw I soon had a roaring log fire going and used its light to put up my igloo snow tent. Next day I drove on and fossicked around the Happy Valley area, exploring boggy side tracks to old mining ruins and motoring off-road here and there to confirm that the current Range Rover will go anywhere the original version would. Which means it can match the performance of any off-the-shelf four-wheel drive — while its occupants enjoy total comfort.

I didn't indulge in any macho-style shenanigens, being content to climb up and down extremely steep, slippery slopes, flounder through deep bog holes, plough across river sand and clamber over submerged mossy boulders while fording various creeks. In low range with the central diff locked the RR does incredible things in third and second gear. I can't imagine ever having to use first, unless I had a heavily laden trailer or a large van in tow. The suspension seems perhaps even softer than before. Which means the vehicle squirms and wriggles in an almost sensuous manner while easing itself over the most difficult terrain.

It's an uncanny experience to enjoy a luxurious ride over rough terrain that

Wild weather in the mountains means there will always be occasional delays due to ice, snow, landslides or fallen trees. Here the resourceful Carter applies his trusty chainsaw to clear the way as the Range Rover heads towards Wainalla

vould have you banging your head and alling about in more basic off-road ve- iicles. The sensation is rather like iding in a sedan chair on the back of an lephant. Very comforting and satisfy- ng, and very distinctive.

At speed on corrugated, pot-holed or uttered surfaces, the Range Rover con- inues luxuriously comfortable, ironing ut the terrain so driver and passengers re left wondering if the surface is eally as rough as it looks. Pushed as ard as is sensible on winding gravel nd earth roads it is astonishingly sure- ooted and merely begins to drift if you verdo things.

At the end of the day it was still rain- ng so I gave up trying to get further cenic photos and camped for the night. returned to Dargo next morning. The ub was still shut. On the winding nar- ow road to Mount Hotham it began nowing and again I was glad of my hainsaw when a fallen tree blocked ny way. It was a narrow, tortuous sec- ion of road cut into a mountainside, vhere the alternative would have been o reverse several kilometres before nding a suitable place to turn. Not a leasant prospect in steadily falling now, with a hike of some forty kilo- netres back to Dargo as the only other ossible alternative if no help came long! The Tramping Club map is cor- ect in stating: "This alpine road is angerous in winter."

When I connected with the main oad to Mount Hotham it began snow- ng in earnest and for a time I had oubts I would get through to Omeo. here was no movement at the ski re- ort and all I saw were several cars arked beside the road, buried in what urned out to be a further 20-centimetre all of snow. With the central diff ocked, the Range Rover ploughed teadily along in second gear high ratio, nly sliding twice when I became too onfident and tried to drive at 25 km/h vhen I should have been doing no nore than 15.

Visibility was at times barely more han 10 metres and I confess I did not njoy the experience or have any incli- ation to stop and take pictures (largely rom fear of being rammed in the back y another idiot who, like me, would ave been better advised not to travel).

The snow eased below Cobungra nd I sped toward Omeo and a late unch of hot pasties beside a pot-bellied tove in a roadside cafe. Not much can e said of Omeo, except that it is not y idea of a destination resort. It does ave some shops, pub and garage. Plus Westpac bank that had never heard of ie Westpac American Express gold

Carter's tour took his Range Rover through some of Australia's roughest country, but it is country which has its own special appeal, not least because it's uncrowded. This is the scene in downtown Woods Point

credit card, despite a sign in their window advertising this facility!

It had been my intention to travel via Benambra and Suggan Buggan down to Buchan and Orbost, before returning home via the Princes Highway. But continuing rain and gloomy forecasts made a return by Albury seem more prudent. Never mind.

The road over the Dividing Range to Mitta Mitta is a beauty for any touring motorist who wants to get away from it all, particularly if he or she likes camping, fishing, tramping, photography, sketching or painting. I had forgotten this, not having been in the area some twenty years. The way is tortuous in parts, but the surface good and no problem for conventional vehicles. There are plenty of side tracks for the off-roader. Mitta-Mitta is an attractive little town at the beginning of the bitumen that takes you finally to the Hume Highway. As a civilised stop-over, it is infinitely more desirable than Omeo.

The sealed run down to Tallangatta didn't take the Range Rover long, despite its meandering course, and was a good test of the vehicle's main-road handling, which I can only describe as faultless, though the slightly disconcerting body squirm takes a little getting used to. The front tyres squeal first and if you continue even harder, there is discernible understeer. By and large, the RR just steams ahead like an express train, with a rather similar body sway at times.

If the vehicle has a fault, I suspect the brakes are just adequate, particularly if you are travelling too fast and are confronted with an urgent reason to stop. Remedy: don't travel too fast! (Do as I say, not as I do.) Petrol consumption? To cover just under 2,500 kilometres, the RR drank 404 litres. This works out on my slate at roughly 6.25 km/l. Not bad for mostly fast travel.

The only thing I didn't like was the plastic front spoiler fitted with driving lights. This serves no useful purpose and would be a damn nuisance in the rough. If ever I manage to save up $58,000 and buy a new Highline (a project I might consider), I would have the contraption off.■

Range Rover: Beauty and the Beast

by Jim Allen

Say "four wheel drive", and you'll likely think of a machine like an old Land Rover; incredibly capable and rugged, but uncompromising in its ability to be uncomfortable. Think of luxury cars, and you'll imagine soft seats, power windows, and air conditioning; hardly something you'd take on safari. This year, Land Rover Ltd. brings to America its most opulent creation, the Range Rover, a vehicle renown for bridging the gap between luxury and four-wheel drive.

A direct descendant of the world-famous Land Rover, Range Rover was conceived as the 'gentleman's' alternative to that old rough-and-tumble four-by. Though first conceived in 1970, this is Range Rover's first try at the American market. Ironically, the designers of Range Rover had the U.S. market in mind even back then, but circumstances prevented them from marketing them here until now. Range Rover of North America, the wholly-owned subsidiary of Land Rover Ltd., forecasts 3000 sales for the 1987 year through its current nationwide, 36-dealer network. The number of dealers within that network will increase to 65 next year and RRONA predicts that unit sales will rise to 5000 by 1989. Sales are aimed at the affluent market, as the $30,825 price tag indicates. The question is, can Range Rover survive in this market as the most expensive sport-utility vehicle available? Range Rover of North America is confident that it can and will.

Charles Hughes, president and chief executive officer of RRONA said recently, "We're bringing to market a totally new concept .a dual-purpose vehicle that can go anywhere in the world - and has - yet retains the comfort and sophistication of a fine European sedan."

The Beauty

Aside from being nice to look at, the Range Rover is comfortable enough to satisfy even the most discriminating drivers. Standard comfort and convenience features rival any luxury sedan from anywhere. The six-way power seats adjust to fit anyone from Willy Shoemaker to Kareem Abdul Jabbar in delicious comfort. At the touch of a button, you can set your cruising speed, open a window, fine tune that right side mirror, set the inside temperature, lock your doors, or clean your halogen headlamps. The sound system, a Clarion NT-2 with four speakers, allows you to drown out city traffic ot enhance your drive in the outback with appropriate mood music.

Your first drive in a Range Rover might feel a little strange. Those of you who are used to low-slung sports cars might get acrophobic at the high seating position. Once you get used to this, however, you'll discover all sorts of advantages, not the least of which is *real* visibility. Range Rover sports a full set of analog gauges that are easy to read and intelligently placed. Controls are also well thought out and lack that 'cheap' feel that British cars were notorious for in days gone by. Standard interior is plush velour, and Connolly leather is optional. In fact, the leather is the *only* option available. Real wood trim inside is a nice touch that warms the interior and gives it a rich feel.

In keeping with the 'four-wheel-drive' aspect of Range Rover, the woven carpets are removable with rubber mats underneath. This allows for the dirtier side of 'luxury four-wheeling'. The Range Rover interior lacks only one item: a decent door armrest. With the obvious effort that went into ergonomics,

This sequence, seen right-to-left, illustrates the Range Rover's enormous suspension compliance. Photos by Jim Allen.

At Home In The 'Range'

by Dave Destler

Comfortable, quiet, and comprehensively-equipped, the Range Rover is all you could want in a car . . . and more.

We had just turned in our new Jaguar XJ6 Vanden Plas roadtest car to Jaguar prior to the new Range Rover being delivered, and we expected an odd initial impression. Since the Jaguar sits relatively low, with the emphasis on luxury, and the Range Rover is a tall, off-road device, we hoped we would be able to be objective about its relative lack of comfort. After all, the Jaguar is a hard act to follow. Well, we got a quick lesson in pre-judging!

The Range Rover's height and road perspective would be alien to a driver who has never owned a van or pickup truck before, but our van has us familiar with the altitude. The comfort was quite something else. The fat padded steering wheel seemed to be in just the right place, while the velour seats, complete with adjustable armrests, were nothing short of pure luxury. (I wondered about Jim Allen's comment about the lack of armrests; My test vehicle had a door-mounted unit that I felt was fine.) With power-adjusted seats, I was able to find a position comfortable for hours of driving. The Jaguar XJ6's seats, although featuring far more positions, failed to provide me with that *perfect* setting I sought. The feel of the Range Rover's seats themselves is plush, but I'm afraid that the velour will look awful in too short a time. Perhaps the leather is worth ordering for this reason, even if it allows a bit more slippage when driving off-road. Headroom is more than enough for anyone, no matter how tall; something else the Jag couldn't boast.

I liked the wood trim on the doors, but feel it should extend across the dash along the natural detail there. This would carry out the 'wood-interior' theme, as in the sedans it emulates. Also, I'd prefer polished wood rather than the natural matte finish. The rest of the interior, although having an 'un-British' look to it, still is very functional, if a bit cluttered. Unfortunately, some of the components have a very cheap look and feel to them, such as the power seat switches, housed in flimsy plastic boxes mounted inboard of the seats. The multi-function warning panel between the gauges looks quite unsophisticated, and the moulded plastic air outlets at the corners of the dashboard also have a low-buck look to them; definitely not to be tolerated in such an expensive vehicle. Fortunately these are easily-remedied shortcomings, not major engineering flaws. The performance of all the various systems was first-rate, however. The main gauges are analog, and so is the clock, I was pleased to see. But why do modern stereos have to be so confusing?

Ride and handling on the road is very good, although obviously not up to the Jaguar XJ6's standards. A trace of uncertainty from the rear axle is evident in quick cornering, but then it was not built to dice with sports cars, and off-road, it's extremely confidence-inspiring. You feel like it can do no wrong there, and has you thinking about moving out to the country so you can drive over bumps on a more regular basis.

The motor performed without as much as a hiccup, although spirited off-the-line performance required really sticking your foot into it. I couldn't help but think what a super candidate it is for a tidy little turbocharger to be feeding those eight cylinders. It did accelerate strongly once under way, partly due to the excellent, responsive four-speed automatic transmission. The motor never ran hot, even in California summer heat with the air conditioning running. In fact, the air conditioning was so good that we never needed to turn it on full blast.

The more we drove it, the more sense the whole concept made. It's fun, versatile, well-constructed, safe, comfortable and exclusive. This last attribute is especially important if you're into one-upmanship. But don't expect to turn heads with a Range Rover. Hardly anyone seems to notice it. Which may be just what you really need, particularly if you have $32,000 to spend on a vehicle that, while obviously the best of its type, fetches almost three times what the similar-in-configuration Isuzu Trooper costs, and six thousand more than America's best Jeep Grand Wagoneer. But if you're in the market for a new luxury car in this price range, and think you might want something that does *more* than a sedan and is built to last forever, I would heartily recommend it.

Even if you never go off-road, it's perfect for going to the market, the golf course, or taking the kids to the park. In fact, I found the Range Rover such a good all-around vehicle that, if I had to own only one vehicle for all my needs, this would probably be my best choice.☐

- Dave Destler

it is surprising that this has gone unnoticed. The driving position lends itself (at least to this writer) to the use of an armrest on long tours and it's annoying to keep reaching for what simply isn't there.

On the road, Range Rover is smooth and quiet. The 3.5 litre, port fuel-injected, aluminum V8 is responsive and torquey. This EFI engine is a vast improvement over the earlier Zenith-Stromberg carbureted engines. Due to its relatively low power-to-weight ratio (150 bhp vs 4300 lb curb weight), acceleration isn't neck-snapping, but adequate. Factory 0-60 time is listed at 13.82 seconds, which is comparable to other vehicles in the sport/utility category, but disappointing considering the price. The 4-speed ZF automatic may partly account for the ho-hum acceleration, but since the five-speed transmissions aren't available in the United States, we won't have the chance to compare. Also, with a larger 3.9 liter engine in the works, we may see an improvement in the power department at a later date. The Range Rover will, however, cruise all day at 80mph, if you want to risk a ticket, and the top speed is listed at a shade over 101mph.

Cornering is excellent, once you get used to the body roll. The compliant, variable-rate coil spring suspension accounts for this body roll and also for the Range Rover's superb off-road performance. With leading and trailing arms, a front Panhard rod, and a rear Boge *Hydromat* leveling device, suspension movement is strictly controlled and precise under all road and load conditions. The standard 205R16 M&S rated Michelins are a vast improvement over the Avon tires offered in previous years, and are a good all-purpose tire. For serious mud or snow, however, a more aggressive (and noisier) tread pattern would be needed.

Braking is superb and far superior to anything this writer has experienced in a four-wheel drive vehicle. With 11.75 inch discs up front and 11.42 inch discs in the rear, there is just over 1100 inches of swept area. The hydraulics are servo-assisted and have four piston calipers up front, two piston in the rear. The discs are solid and no fade was noticed after repeated stops. The Range Rovers tested had one annoying habit related to the brakes; a terrible screeching when the brakes were applied in reverse. In all fairness, this could be due to the tremendous abuse these test vehicles have seen at the hands of journalists who wanted to give them a good checking out.

Continues page 85

SPECIFICATIONS: 1987 RANGE ROVER

ENGINE	
Type	Aluminum V8, Dry steel liners
Displacement	3528cc (3.5L), 215 cid
Bore X Stroke	88.9mm X 71.14mm (3.5in X 2.8in)
Compression Ratio	8.13:1
Maximum SAE Net Power	150bhp @ 4750 RPM
Maximum torque	196 lbs/ft @ 3000 RPM
Fuel System	Lucas electronic fuel injection
TRANSMISSION	
Gearbox	ZF 4-speed auto w/lockup torque converter
Transfer Box	2-speed, lockable differential
Axle Ratios	3.54:1
SUSPENSION/CHASSIS	
Front	Coil springs, radius arms, Panhard rod
Rear	Coil springs, Boge Hydromat leveling device
Travel	Front: 8 inches, Rear: 11 inches
Steering	Power-assisted worm and roller
Wheels	Alloy, 5-stud 7 X 16 inches
Tires	Michelin X 205R16 'MS 200'
Brakes	4-wheel discs, 4-piston front, 2-piston rear
Chassis Type	14 gauge boxed steel ladder type
Body	Aluminum outer panels; steel hood, tailgate, rear lower quarter panels and underbody
DIMENSIONS	
Overall Length	175 inches
Overall Width	71.6 inches
Overall height	70.8 inches
Ground Clearance	7.5 inches
Curb Weight	4303 lbs.
Wheelbase	100 inches
Track	58.5 inches
Fuel Capacity	21 gallons
PERFORMANCE	
0-30 mph	4.37 seconds
0-60 mph	13.82 seconds
Standing ¼ mile	19.6 seconds @ 70.4 mph
Maximum Speed	101.4 mph
Fuel Economy (EPA average)	14 mpg
Turning Circle	39 feet
Approach Angle	30 degrees w/spoiler, 42 degrees w/o
Departure Angle	30 degrees
Ramp Breakover Angle	29 degrees
PRICE	
Suggested Retail	$30,825
Optional Leather Upholstery	$1,025
Transportation	$550
Warranty	3 Years, 36,000 miles full vehicle; 6 Years, unlimited mileage corrosion perforation warranty

PHOTO BY DON HUNSTEIN

RANGE ROVER

Surviving Manhattan, or, Clive goes native

BY TED WEST

"What to do, Penelope, what to do! The civilized world is crumbling! The only really civilized city in America, as any cultured person knows, is New York City . . . and New York City is an unmitigated shambles. The subways, Broadway, the ethics of municipal government, the roads—especially the roads. They've all been reduced to rubble. Ah, Penelope, what to do!"

"Come along, Clive," Penelope smiled.

"Step into my Range Rover and we'll make it all well."

INDEED, HOW DOES a cultured person weather the steady decline of the American Rome? (He moves to Topeka and starts enjoying life, you say—but that's because you're simply not a cultured person like our Clive.)

Well, at least one solution to this quandary is provided by Clive's beloved—and *decidedly* cultured—sister Penelope. Her Range Rover may not improve standards in the New York theatre or amend the infinite misdeeds of Mayor Koch's nefarious underlings—but one thing it'll do for sure, it'll mend the roads.

Let's ride with Clive, then, and see what he learns. . . .

The first quality he notes upon approaching dear Penelope's 4-square 4-wheeler is the dignified demeanor of the thing. Spot on, thinks he. No mistaking this for some awkward, re-coutured Yankee truck. It has about it an undeniable stiff-upper-chin, damn-the-tornados/4-speed-ahead staunchness. He's convinced. Not only would he be quite at home crossing the African veldt in such a vehicle, he'd be quite at home sipping fine port and talking bully times while doing so. Crossing Manhattan, of course—a more demanding trek—he'll require a bot-

tle of suitably profligate "designer water" (2/$7.00, on sale).

But that can be arranged. And having sized up the Range Rover's handsome leather interior and honest-to-God, you-could-burn-it-in-the-fireplace walnut trim, he's sure the Range Rover's well-bred dignity will be invulnerable to Gotham's inherently uncouth nature. Even in New York, the easy discussion of bully times will be neither strained nor brief in this Contemporary Victorian.

His continuing inspection further confirms the Range Rover's gentlemanly virtues. It's got everything a civilized city-trekker requires—four doors, air conditioning, excellent stereo, fully adjustable power seats with Mercedes-Benz-style analog adjuster, electric window lifts and a rear window "child block" master switch, analog tach and speedometer, central door locking, a cluster of 14 dashboard warning lights that'll tell him if his shoelaces are untied, 40-cu-ft storage capacity, a Class 3 tow hitch to haul up to 7700 lb. Much of the Range Rover's body is aluminum, making rust the rarest of luxuries, and it's even got "puddle lamps" in the base of the doors so that at night you can see what you've stepped in.

Climbing aboard, Clive notes the nice short-throw shifter for the 4-speed automatic. He pumps the pedal of the 4-wheel disc brakes and finds nothing amiss. As the Range Rover is endowed with full-time 4-wheel drive, there will be no crawling through the mud to lock hubs—though Clive may be obliged to operate the ultrashort auxiliary shifter from time to time. This worthy device serves both to select high or low range (providing a total of eight drive ratios) and to lock or unlock the center differential. It operates a bit stiffly until you get the hang of it, but makes an amusing little *peep!* each time you successfully change ranges.

The list of automotive amenities, then, is nothing if not generous, especially for a 4-wheel-driver. So far, so good. The R-R (exceedingly important initials to our Clive) has passed the Pedigree Test.

But what looms now is a more grim affair, the Loose-In-The-Streets Test. Therefore, for reasons having nothing whatever to do with recent New York legislation, Clive solemnly buckles up. He tells Penelope to do likewise. He even removes his beige-tinted designer glasses (worn for effect only—remember, we're talking Culture here).

The first thing Clive notes as he pulls away from the fire hydrant is that, compared with other RVs, the Range Rover has an exceptionally high greenhouse. This is useful not only for seeing out but, equally important to Clive, for being *seen in*. And, though one might not suspect it from the outside, the Range Rover is surprisingly compact compared with most such vehicles, making it a far more difficult target for rogue taxis and territorially enraged delivery vans.

Further contributing to the Range Rover's excellent natural cover in the field, its sharp-edged fender limits make for extremely accurate and confident maneuvering in tight quarters, particularly when vying for position at a stoplight. Indeed, early on at a stoplight, Clive came very close to an *affaire d'honneur* with a surly fellow in a green-and-rust '73 Grand Prix. When Clive proposed rapiers at dawn, however, the other, thinking there was something sexually wrong with our man, moved on.

But let loose once again on the Manhattan torrent, Clive and the Range Rover were left to their own devices. And primary among these devices, of course, was the 3.5-liter aluminum Rover V-8. It generates 150 bhp and 195 lb-ft of torque, providing, well, "adequate" thrust for this 4300-lb vehicle. Here at last Clive found a reason to whinny. The Rover's power may be plenty for trekking across Uganda dragging rhino carcasses, but for dragging New York yahoos to the next empty parking space, it's definitely second best, a clear case of *Un*-Sudden Acceleration. What the engine lacked in quickness, however, it made up for in satiny V-8 smoothness.

Contributing inordinately to the Range Rover's lackluster performance in the midtown cut-and-thrust, it should be added, was the transmission. When Clive tromped on the noise pedal, it seemed the automatic took an unseemly amount of time shuffling pages, before going through its kick-down.

But once the transmission made its decision, the engine came up to speed and Clive and Penelope were launched through the Lower Manhattan veldt, that's when the Range Rover came into its own. Whatever else may be said of this vehicle, its ability to apologize for potholes and bomb craters is utterly unmatched. The 4-wheel coil spring live-axle suspension, with its perfectly calibrated rising rate springs, seems to start out just right and, as things get craggier and more offensive, *stays* just right. In fact, accelerating from zero-to-hellacious through the worst mid-street mountain ranges Manhattan has to offer, Clive was able to keep his glass of imported status-water three-fourths full and not spill a bubble.

Of course, as Clive also noted to the long-suffering Penelope, with all this plush-riding luxury comes a price. Turning corners vigorously or avoiding stubborn mid-crosswalk pedestrians tends to pavement-polish the door handles perhaps overmuch. In short, body roll is in the QEII-Meets-Hurricane-Lorna range. Most will become used to this motion soon enough, of course. And for the faint of stomach, there's always a healthy dose of Dramamine.

However, a Range Rover quality that took a bit more getting used to was the vehicle's steering response. Thanks, no doubt, to the full-time 4-wheel drive, the Range Rover has relatively little caster effect, or self-centering. This means Clive had not only to steer the vehicle into a turn, but also resolutely steer it back to straight ahead. And, alas, Clive is a somewhat absent-minded fellow, prompting poor Penelope to screech at the top of her charming lungs on more than one occasion.

Furthermore, in the turns Clive sensed a sort of rubber bandish sensation, as if the Range Rover were swiveling around its own maneuver. Thanks again, no doubt, to its unique suspension system, the Range Rover has a noticeable degree of compliance steer, as if making all those heavy, driven pieces turn this way and that is causing the entire suspension locating system to flex. This isn't a serious matter, just a bit distracting.

The afternoon progressed apace and, in the end, no damage was done—nor, might we add, did it seem likely much damage *could* be done to this tougher-than-tough critter. Indeed, it wasn't long before Clive was making wild swerves across multiple lanes of mid-Manhattan blacktop just to get to the nearest brutal crater up ahead. The Range Rover would approach said cavity, stare it in the eye, swallow it down like a sweet pill and ramble on at full speed toward the next chasm on the menu . . . and still there wasn't a drop of imported water to be found on the plush gray carpet.

Finally, Clive and Penelope returned home. Penelope wasn't talking much now. She complained of hoarseness.

Clive, on the other hand, was beside himself with good spirits and bully talk. Except for a recurrent soft moan at the Range Rover's left rear, as if a thirsty suspension bushing were quietly begging a jigger of chassis lube, this worthy vehicle remained just as tight and silent as it'd been at the start. Clive was sold. He needed only to know the price.

"$31,000," Penelope croaked weakly, fearing his reaction.

But she needn't have worried, Clive was a true New Yorker. "Why that's exorbitant!" he beamed immediately, "I'll take it!"

And they walked off to dinner, bewailing the sad state of the theatre.

PRICE

List price, all POE	$30,825
Price as tested	$30,825

Price as tested includes std equip: air cond, AM/FM stereo/cassette, elect. window lifts, elect. adj seats, elect. adj mirrors, cruise control, central locking

GENERAL

Curb weight, lb	4315
Test weight	4445
Weight distribution (with driver), f/r, %	49/51
Wheelbase, in.	100.0
Track, f/r	58.5/58.5
Length	175.0
Width	71.6
Height	70.8
Trunk space, cu ft	28.2 + 11.7
Fuel capacity, U.S. gal.	20.0

ENGINE

Type	ohv V-8
Bore x stroke, mm	88.9 x 71.1
Displacement, cc	3532
Compression ratio	8.1:1
Bhp @ rpm, SAE net	150 @ 4750
Torque @ rpm, lb-ft	195 @ 3000
Fuel delivery	elect. port inj
Fuel requirement	unleaded, 87 pump oct

Exhaust-emission control equipment: 3-way catalytic converter, EGR

DRIVETRAIN

Transmission..... automatic, torque converter with 4-sp planetary gearbox

Gear ratios: 4th (0.89)	3.84:1
3rd (1.21)	5.23:1
2nd (1.81)	7.82:1
1st (3.02)	13.04:1
Transfer ratio x final drive ratio	1.22 x 3.54

CHASSIS & BODY

Layout	front engine/4wd
Body/frame	steel & aluminum/steel
Brake system, f/r	11.8-in. discs/11.4-in. discs, vacuum-assisted
Wheels	alloy, 16 x 7J
Tires	Michelin X M+S 200, 205R-16
Steering type	recirculating ball, power-assisted
Turns, lock-to-lock	3.5
Turning circle, ft	39.0

Suspension, f/r: live axle on leading arms, Panhard rod, coil springs over tube shocks/live axle on lower trailing arms, upper trailing A-arms, separate coil springs and tube shocks; automatic ride leveling

CALCULATED DATA

Lb/bhp (test weight)	29.6
Bhp/liter	42.5
Engine revs @ 60 mph in top gear	2340
R&T steering index	1.37
Brake swept area, sq in./ton	222

ROAD TEST RESULTS

ACCELERATION

Time to distance, sec:	
0–100 ft	3.9
0–500 ft	10.5
0–1320 ft (¼ mi)	19.0
Speed at end of ¼ mi, mph	72.5
Time to speed, sec:	
0–30 mph	4.1
0–40 mph	6.3
0–50 mph	9.2
0–60 mph	12.6
0–70 mph	17.4
0–80 mph	26.3

SPEEDS IN GEARS

Maximum engine rpm	5200
Rpm limiter	5200
4th gear (rpm) mph, est	(2560) 90
3rd, est	(4470) 90
2nd	70
1st	43

FUEL ECONOMY

Normal driving, mpg	13.5

BRAKES

Minimum stopping distances, ft:	
From 60 mph	179
From 80 mph	316
Control in panic stop	fair
Overall brake rating	poor

HANDLING

Lateral accel, 100-ft radius, g	0.66
Speed thru 700-ft slalom, mph	55.0

INTERIOR NOISE

Idle in neutral, dBA	52
Maximum, 1st gear	74
Constant 70 mph	71

ACCELERATION

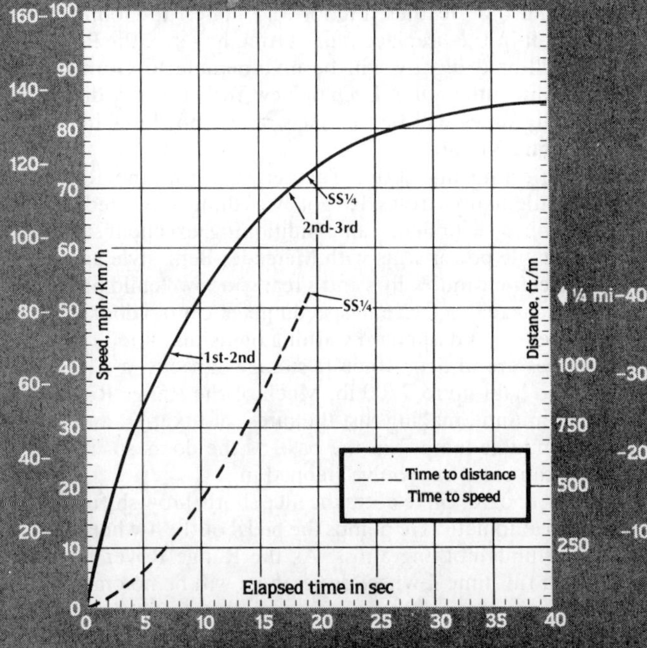

Range Rover Vogue EFi

The rest of the civilised off-roaders are catching up. Models such as the Mitsubishi Shogun and Isuzu Trooper now make a good fist of the job which used to fall only to the Range Rover, that of combining toughness and traction off-road with an ability to cruise quietly home on the motorway at 90mph. This, the marketeers tell us, is very much the way things are going in the 4x4 market.

Still, the Shoguns and the Troopers haven't yet made the grade, we reckon. The look and the performance of the Range Rover still eclipses the best of them, which makes the fact that it is 18years-old seem the more remarkable.

The Range Rover's special appeal has always been founded on two main things: the superb style and design of its body, and the special qualities of its all-aluminium 3.5litre V8 engine. David Bache designed its body, as he did the shape of the Rover SD1. It seems tragic, now, that he should have been a victim of BL Cars design department infighting, The original two-door Range Rover body had a combination of style and utility about it which makes decade-old examples still look good to modern eyes.

But the body design has another big virtue. The Range Rover carries over the Land Rover policy of using aluminium for most of its panels, a fact that means a high proportion of hard-used early examples still ply the roads of Britain. But there are a couple of disadvantages. One is that the long life of the alloy panels highlights the tendency of the Range Rover's tailgate, built of steel, to corrode very smartly indeed. The other is the fact that aluminium panels are more difficult to form than those in steel. So Range Rovers have large and unsightly panel gaps (though, admittedly, they're also designed in, to allow for body flex on difficult terrain), and their surfaces are more prone to rippling.

'The V8 has no peer for smoothness or for discreet torque delivery'

A big reason for the Range Rover's rule-breaking refinement is the sophistication of its big, rubber body bushes which isolate the main structure from its tough, box section chassis. Such trouble had simply not been taken with such machines before 1968. Also, the decision to suspend the Range Rover's non-independent live axles, front and rear, on long-travel coil springs – with carefully calibrated damping – gave it a sophistication that previous off-roaders, which almost all used semi-elliptic springs, could not dream of matching.

Yet for all its original promise, the Range Rover's more recent progress is what puts it in this year's *Top 10*. It was built for more than a decade without four doors, or a five-speed gearbox to give it a true motorway cruising ratio, or even the option of an automatic transmission (it now has the excellent ZF four-speeder). And only recently did it acquire the equipment which is routine in many machines of half its £20,000 cost – central locking, a decent stereo, electric windows and a properly plush interior. These now all go into the best-seller, the £21,192 Range Rover Vogue EFi.

This machine, the one we feature here, comes equipped with the recently uprated 165bhp fuel-injected version of the V8 (107mph, 12.0sec 0-60). There's also a 112bhp VM turbodiesel (90mph, 14.6sec 0-60). Neither model's performance details seem impressive by high-power saloon standards, yet they translate into brisk performance on the road – 100mph cruising in the V8, if needed, and quite enough go from the diesel, too, to make it an all-day, good-roads proposition.

Range Rovers make plenty of sense, then, but it is the enjoyment to be had from driving them which makes them exceptional. The V8 engine has no peer among the pretenders, for smoothness, discreet torque delivery or the ability to amble along at low revs on a whiff of throttle – an advantage both in the suburbs and the bush. Coupled with the optional four-speed autobox, this is a luxury saloon's power train.

And if all this still isn't enough for you, there is the suspension, so superior to the rest of them – better damped, more comfortable, quieter and able to provide handling and high speed manoeuvrability which logic says is inconsistent in so tall a machine on cart axles.

There are drawbacks, of course. Range Rovers are not as well made as they might be, a fact partly explained by their aluminium outer body panels. And the V8's fuel consumption (an 18mpg average is as good as a normal driver must ever expect) will probably be considered crippling by many.

Yet the message is as it has been for almost 20years. The Range Rover remains the best and most stylish all-purpose vehicle in the world. Also the most enjoyable to drive and, if you care, the best to be seen in. Sure, the forces of opposition are in sight, and better priced, but you couldn't honestly say they were in the overtaking lane.

AT-A-GLANCE

FOR

Lively performance, comfortable cabin, fine ride, secure and fluent handling. Excellent automatic box

AGAINST

Heavy fuel consumption, initial purchase price

PRICE	*£22,315*
MAX SPEED	*102.5 mph*
0-60 MPH	*12.3 sec*
MPG OVERALL	*16.4*
TEST DISTANCE	*1618 miles*

It is a measure of the brilliance of the original Gordon Bashford/Spen King design that, 18 years after its introduction, the Range Rover is still the yardstick by which other all-terrain vehicles are judged. More than that, there is no single area – apart from fuel consumption – in which the high standards it sets are even approached.

This British success story is even more remarkable because, for the first 10 years of its life, the Range Rover was starved of development. Since the appearance of the four-door version in 1981, however, Land Rover Ltd have carried out a long-overdue programme of diversification and refinement that has enhanced both the performance and the image of a car that has always been notable for its charisma.

An automatic gearbox was first offered in 1982, coupled to a redesigned two-speed transfer box/centre differential unit that virtually eliminated the gear whine that had been one of the car's few demerits. The loss of performance inseparable from a conventional automatic box was remedied four years later when the Range Rover's magnificent light-alloy V8 3528 cc engine was equipped with fuel injection.

Our test car was the latest of the automatics – the luxury-specification Vogue, with ZF four-speed box and fuel-injected engine. The automatic option adds £1123,75 to the basic Vogue's £21.191.63 price tag – a figure that includes, as standard equipment, exterior, interior and electrical packs that are £2664.83 options on other models in the range. Only a few options are listed for the Vogue itself, with air conditioning (£1345.50) the most expensive. The others are a towing pack (£105.89), heavy-duty front suspension (£43.61) and – oddly – front mud flaps at £24.92. Our test car also boasted an electrically-operated sun-roof at £850.91.

A "full house" Vogue could, therefore, cost upwards of £23,000. For that sort of money, one could buy a BMW M535i, a Mercedes 300TE, a Porsche 944, a Saab 9000 Turbo 16SE, or a massive Volvo 760 Turbo Estate. All offer far more performance, while lacking the Range Rover's versatility. But, the same expenditure would cover a perfectably capable though less charismatic Japanese all-terrain 4wd *and* a Car of the Year Peugeot 405 SRi.

Is the Range Rover Vogue really so good that it can beat competition like that? Step up into its lush limousine interior, hear the throb of its so-smooth V8, savour the security of permanent four-wheel drive, enjoy the exalted view from your high perch, looking over the rooflines of surrounding traffic, and you will have few doubts. Above all, the Range Rover offers a blend of luxury and practicality that simply places it in a class of its own. It is one of the few cars of which it can truly be said that anything else is less.

It imposes itself on the eye from the start. It is a *big* car, but its lines have a classic simplicity that understates the fact. Some subtle reworking in recent years – the addition of restrained wing bulges to make width judgement easier; the substitution of concealed bonnet hinges for the external type that always jarred one's sensibilities – have made it even smoother than before. The front air dam is restrained in its proportions, and offers a neat built-in housing for a pair of fog lamps. This is a car clearly conceived as an entity. It has style and character.

Internally, in Vogue form, it has luxury too. The driver and passenger can enjoy, literally, armchair seating. Cloth upholstered, with an ample range of fore-and-aft adjustment and reclined backs, the front seats are simply magnificent – though one tester felt that he could still not obtain real under-thigh support. The split rear bench is almost equally comfortable, with ample legroom and – of course – a folding centre arm rest. The built-in rear head restraints are effective, but set low enough not to obstruct the driver's rearward vision when reversing.

Stowage is less generous. Each front door has a modest bin and there is space for bulkier items in the centre console. No glovebox, though: its place is partly taken by a neatly concealed and very well-placed fuse box.

Who really needs to worry overmuch about oddment stowage, though, when the Vogue has such a massive general carrying capacity? The boot space – now concealed beneath an elegant parcel shelf – is a huge 36.18 cu ft, despite the presence of the 16 in spare wheel. With the seats lowered, the cargo space increases to 70.8 cu ft. A welcome detail improvement on current models is that the tail-gate release lever is now internal, and a pair of grips has been provided on the upper edge of the gate so that hands no longer have to grasp a dirt-besmirched exterior.

Few cars can match the Range Rover for the lightness and airiness of the cabin, or for its unobstructed field of vision. And the driving position is virtually perfect, marred only by a gear selector knob that is so wide that it is only too easy to trap one's hand – painfully – when operating the handbrake at traffic stops. All the other controls are so well positioned that one almost suspects that this is a deliberate fault, on the analogy of Persian carpet makers incorporating one defect to

;uard against blasphemy!

Range Rover drivers have nore reason than their exalted)erch for feeling on top of the vorld, however, because this s, dynamically, a surprising :ar. It is easy to be misled, by ts height, into thinking that the iandling could be suspect. Nothing could be further from he truth, for the car is built on heavy and rigid chassis, and he main engine/transmission veight is low-set. The body it-elf is comparatively light, and he combination of permanent our-wheel drive and a low cen-re of gravity provides the Range Rover with excellent oadholding and an unexpected degree of agility. Under sharp ornering, there is very notice-ble body roll, but the car holds ts line and grips tenaciously. It an be cornered as fast as any onventional estate car, right p to a limit whose approach is otified by the onset of front xle tramp. At that point, it will ose its dignity and lurch into afe understeer – but it takes lmost conscious mishandling o provoke it even to that de-;ree of protest.

One is always aware, of ourse, that this is a car with wo live axles. Transverse rip-les can be felt through the teering, and on modestly ough surfaces there is a de-ree of bump steer and slight ggling of the front axle. Really roken surfaces produce no uch effects: the worse the terrain beneath the big 16 in wheels, the better the car handles and rides. It is, after all, designed to tackle heavy off-road going that would bring most other cars to a sudden and ignominious stop.

The steering, too, is necessarily biased towards the Vogue's off-road role. While surprisingly accurate for a worm and roller system, it tends to be on the light side on metalled roads, with a certain nervousness around the straight-ahead position. It takes a few miles to develop the requisite delicacy of touch, but once it is acquired a competent driver can hustle the Vogue along winding country roads, flowing through the twists and turns with real poise and composure. Off-road, picking a way through mud and rocks is reduced to mere wheel-twiddling. The most daunting hazards seem to require little more of the driver than a delicate foot on the throttle and selecting the right line. The car itself looks after the rest.

Such characteristics could hardly fail to produce a good ride, no matter what the surface beneath the wheels may be. On tarmac, only sharp-edged pot-holes produce any suspension harshness. Everything else is soaked up by the Vogue's long-travel suspension and, though some lateral movement is noticeable, there is no discernible pitching.

Performance is perhaps the last consideration where potential or actual Range Rover owners are concerned. There are two ways of looking at it. Either, here is a 3.5-litre car that can be comfortably out-run by virtually any modern 1.6-litre supermini; or, here is a luxuriously equipped five-seater go-anywhere touring car that can hold its own even against the liveliest "under 1.4-litre cars" that comprise much of the average traffic stream.

What is undeniable is that both speed and acceleration have benefited greatly from the adoption of fuel injection, which has boosted the V8's power output by 38 bhp and its torque – already massive – by 13 lb ft. The result, against the watch, is dramatic. Compared with the carburettered five-speed manual Range Rover we tested in 1983, the better-equipped Vogue is a full 7 mph faster and, at 12.3 sec, some 2.3 sec quicker in the benchmark 0-60 mph sprint. Fuel consumption has improved from 15mpg to a healthier 16.4 mpg overall. This, despite the power-absorption penalty that use of an automatic transmission usually entails.

To the driver, the Vogue's extra performance potential is not immediately obvious, and on our test car it was marred by a tendency for the engine to sulk, and refuse to pull, in the first few miles from a cold start. On several occasions, it almost died when halfway across a busy road. Once it had warmed, however, the long throttle pedal movement – it is between six and seven inches – also gave an impression of lethargy. It soon became apparent, though, that to obtain the best response from this particular engine/transmission pairing it was essential to use the last three inches of throttle travel, since in the interests of economy the ZF gearbox has been programmed to give an early upchange. A tentative dab on the pedal, therefore, fails to bring in the kickdown. Flooring the pedal, by contrast, gives a fast changedown – perhaps a little too abrupt for smooth progress – and a welcome surge of power.

Used from low speeds, kickdown produces acceleration strong enough to surprise other drivers. Keep the pedal to the floor and the ZF changes almost imperceptibly from one gear to the next, with only slight driveline vibration noticeable. One tester reported a modicum of backlash with lock-up Drive engaged, but in general the ZF 'box won general approval. It suits the engine's characteristics well, gives relaxed road driving, and makes off-road excursions easier providing the centre differential is locked before really difficult terrain is tackled.

Apart from a slightly mushy feel to the pedal, the Vogue's brakes proved to be tremendously powerful yet progressive in action. On the test car, though, the handbrake was badly adjusted and incapable of holding the Vogue even on a slight incline. This was no problem, of course, if facing uphill: the car was then prevented from rolling backwards by the usual sprag in the transmission. Facing downhill, however, the driver either had to keep the handbrake on by manual pressure or hold the car on the footbrake.

It is not simply the more powerful engine and the new transmission that have given the Vogue extra refinement. Detail body improvements have made it a quieter and more pleasant high-speed cruiser than its predecessors, particularly by eliminating wind noise. This, and the parallel reduction in transfer box noise, have produced a much improved driving environment, with interior noise levels down by 3 dBA at a 70 mph cruise and by 5 dBA on maximum readings. Not luxury saloon standards, by any means, but at least now comparable to the noise levels produced by equivalent load-carriers such as the Renault 21 Savanna.

Range Rover Vogue

Still on top, 18 years after its introduction, Britain's heavyweight all-terrain estate can take on all-comers in its latest luxury form – on road and off

Photographs by Peter Burn

PERFORMANCE

WEATHER CONDITIONS

Wind	4 mph
Temperature	37 deg F/3 deg C
Barometer	30.1 in Hg/1020 mbar
Surface	Dry tarmacadam

MAXIMUM SPEEDS

	mph	kph
Banked circuit (3rd gear)	102.5	165.0
Max speed in 4th gear	101.6	163.5
Terminal speeds:		
at ¼ mile	74	119
at kilometre	90	145
Speeds in gears (at 5000 rpm):		
1st	39	63
2nd	66	105

ACCELERATION FROM REST

mph	sec	kph	sec
0-30	4.1	0-40	3.3
0-40	6.2	0-60	5.5
0-50	9.1	0-80	9.0
0-60	12.3	0-100	13.2
0-70	17.2	0-120	20.3
0-80	24.0	0-140	30.8
0-90	34.4	0-160	43.1
Stand'g ¼	18.9	Stand'g km	34.8

ACCELERATION IN KICKDOWN

mph	sec	kph	sec
20-40	3.7	40-60	2.2
30-50	5.0	60-80	3.6
40-60	6.2	80-100	4.2
50-70	8.1	100-120	7.2
60-80	11.7	120-140	10.4
70-90	17.2		

FUEL CONSUMPTION

Overall	16.4 mpg
Touring*	18.9 mpg
Govt tests	14.6 mpg (urban)
	26.2 mpg (56 mph)
	20.2 mpg (75 mph)
Fuel grade	97 octane 4-star rating
Tank capacity	17.5 galls 79.5 litres
Max range*	330 miles
Test distance	1618 miles

*Based on official fuel economy figures – 50 per cent of urban cycle, plus 25 per cent of each of 56/75 mph consumptions.

STEERING

Turning circle	38 ft 10 in	11.8 m
Lock to lock	3.75 turns	
Turning circle for one turn of lock	70 ft 0 in	21 m

NOISE

	dBA
30 mph	64
50 mph	70
70 mph	75
Maximum†	78

†Peak noise level under full-throttle acceleration in 2nd

SPEEDOMETER (MPH)

True mph	30	40	50	60	70	80	90
Speedo	31	41	51	62	72	83	95

Distance recorder: Accurate

WEIGHT

	cwt	kg
Unladen weight*	36.8	1870
Weight as tested	40.9	2080

*No fuel

Performance tests carried out at 2509 miles by *Motor's* staff at the Motor Industry Research Association proving ground, Lindley, and Millbrook proving ground, near Ampthill.

Test Data: World Copyright reserved. No reproduction in whole or part without written permission.

SPECIFICATION

ENGINE

Designation	—
Cylinders	V8
Capacity	3528 cc
Bore/stroke	88.9/71.1 mm
Max power	165 bhp (123 kW) at 4750 rpm (DIN)
Max torque	207 lb ft (280 Nm) at 3200 rpm (DIN)
Block	Aluminium alloy
Head	Aluminium alloy
Valve gear	Push-rod ohv, 2 valves per cylinder
Compression	9.35:1
Fuel system	Electronic fuel injection (Lucas)
Ignition	Breakerless
Main bearings	Five

TRANSMISSION

Layout	Longitudinal front engine, permanent four-wheel drive
Type	Four-speed ZF automatic
Internal ratios and mph/1000 rpm	
Top	0.73/26.5
3rd	1.00/19.4
2nd	1.48/13.1
1st	2.48/7.8
Rev	2.09
Transfer box	1.22 (high), 3.32 (low)
Final drive	3.54

AERODYNAMICS

Cd 0.47	CdA 1.22

SUSPENSION

Front	Live axle located by Panhard rod and radius arms; coil springs and telescopic dampers
Rear	Live axle located by central A-frame, trailing links. Boge-Hydromat ride-level unit, coil springs and telescopic dampers

STEERING

Type	Worm and roller
Assistance	Yes

BRAKES

Front	Discs, 11.8 in dia
Rear	Discs, 11.4 in dia
Servo	Yes
Circuit	Split front/rear
Handbrake	Drum, operating on rear transmission

WHEELS/TYRES

Type	Alloy, 7 × 16 in dia
Tyres – size	200/205 R16
make	Michelin XM+S
Pressures F/R (all conditions)	28/35 psi 1.9/2.4 bar

GUARANTEE

Duration	12 months, unlimited
Rust warranty	None

MAINTENANCE

Major service	12,000 miles
Intermediate	6000 miles
Oil change	6000 miles

STANDARD EQUIPMENT

Seat back map pockets	■
Split folding rear seat	■
Courtesy light delay	■
Map reading light	
Boot light	■
Central door locking	■
Remote boot/hatch release	
Remote fuel flap release	■
Remote control door locking	
Electric mirror adjustment	■
Heating mirrors	■
Intermittent wipe	
Programmed wash/wipe	■
Headlamp wash/wipe	■
Internal headlamp levelling	
Driving lamps	■
Fog lamps	
Radio/cassette player	■
Electric aerial	‡
Electric sunroof	□
Illuminated vanity mirror	
Seat height adjustment (driver)	
Seat tilt adjustment (driver)	■
Seat lumbar adjustment driver/passenger	
Electric seat adjustment driver/passenger	
Heated seat driver/passenger	
Adjustable upper seatbelt mounting	■
Adjustable steering column	
Rear seat head restraints	■
Rear centre armrest	■
Anti-lock braking system	
Rear compartment heating	■
Tinted glass	■
Air conditioning	□
Power assisted steering	■
Self levelling suspension	■
Cruise control	
Limited slip differential	
Leather trim	

‡ Incorporated in rear screen
□ Option

COSTS AND SERVICE

Insurance group	7
Major service interval, miles	12,000
Intermediate service, miles	6000
Oil change, miles	6000
Set brake pads (front) £	45.29
Complete clutch £	134.07
Complete exhaust £	340.62
Front wing panel £	155.25
Oil filter £	9.82
Starter motor £	160.80
Windscreen £	126.56
Tyre £*	77.05
Total service time, hrs – up to 50,000 miles	30.9
Time to change clutch, hrs	5.9

* Discounted price

Above: Dashboard-mounted fuse box. Below: Overwide gear lever knob can trap fingers.

THE WORKS

More than 30 years of development is reflected in the Vogue's all-alloy 90 deg V8 engine – a notably oversquare design (88.9/71.12 mm) – with its power output boosted from 12[5] bhp at 4000 rpm to 165 bhp a[t] 4750 rpm by the adoption o[f] fuel injection. Push-rod valve actuation through self-adjusting hydraulic tappets is re[tained]. Transmission is via a Z[F] four-speed automatic gearbox with overdrive top, and a two speed transfer box incorporating a lockable centre differential. Drive is taken to all fou[r] wheels through front and re[ar]

Left: Armchair seating for driver and front passenger. Above: Clear instruments and warning light bank. Right: The driving compartment is an object lesson in neat, functional design

Far left: Kneeroom in the rear is ample and the seating comfortable, with plenty of headroom. Left: Even with the seats raised, boot space is enormous. The tailgate now has an internal release handle Right: Screenwiper pattern leaves few blind spots for a high-seated driver

Left: Split rear seats make the load space even more versatile. The carrying capacity is enormous Right: The magnificent 3.5 litre all-alloy V8 engine has gained a new dimension thanks to fuel injection

propshafts to live front and rear axles equipped with spiral bevel differentials. The axles are located by radius arms and trailing links respectively. The front axle has a Panhard rod, and the rear axle a central A-bracket. Coil springs and telescopic dampers are used on the long-travel suspension system, with a Boge-Hydromat self-levelling strut at the rear.

Specially-finished heavy-duty light alloy wheels are fitted with Michelin XM+S 200 205 R16 dual-purpose radial tyres.

VERDICT

At the time of its conception, in the mid-'60s, the Range Rover was unique – and it remains so today. Although many all-terrain vehicles have been produced throughout the world in the intervening years, none has yet appeared that can challenge the supremacy of the Range Rover. Especially in its luxury variant – the Vogue automatic – it simply remains in a class of its own. The Vogue's blend of superbly comfortable accommodation for up to five passengers, 100 mph-plus top speed, adequate acceleration, the security of permanent four-wheel drive, and its massive off-road ability offset its two main disadvantages of high initial cost and relatively heavy fuel consumption.

The proven durability of its construction linked to its smooth, low-stressed 3.5-litre engine make the Vogue an attractive long-term proposition – only the number plate betrays a Range Rover's age at a casual glance.

It simply has to be the best buy in its market sector. M

MOTOR'S RATING

■■■■■■■■□□

VALUE RATING

■■■■■■■□□□

Make: *Range Rover*
Model: *Vogue*
Made by: *Land Rover Ltd*
At: *Solihull*
Sold in the UK by: *Land Rover Ltd, Lode Lane, Solihull B92 8NW*
Tel: *021-722 2424*
Number of dealers: *400*
Total price: *£21,192*
Extras fitted to test car: *Automatic gearbox £1124, electric tilt/slide sunroof £851, towing pack £106*
Other options: *Air conditioning £1345, front mud flaps £25*
Price as tested: *£23,272*

Other possible rivals include the Isuzu Trooper Citation 2.6iLWB (£14,998) and Mitsubishi Shogun Diamond 2600 (£14,569

RANGE ROVER VOGUE AUTO £22,315

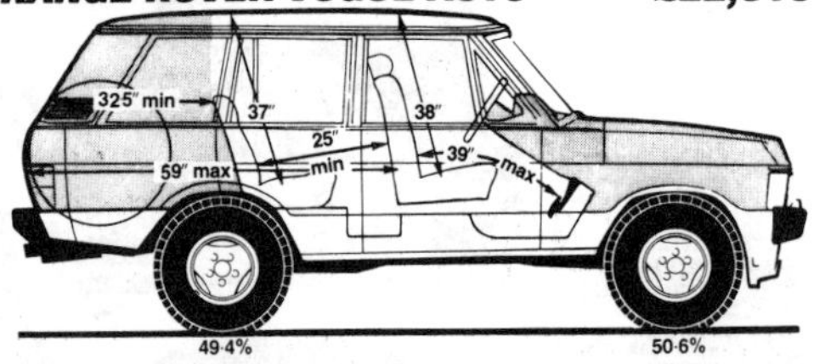

49·4% 50·6%

Length 176" Width 70" Front track 58·5"
Wheelbase 102" Height 70" Rear track 58·5"
Width between w/a 43·5" Floor height 40" max 33·3" min

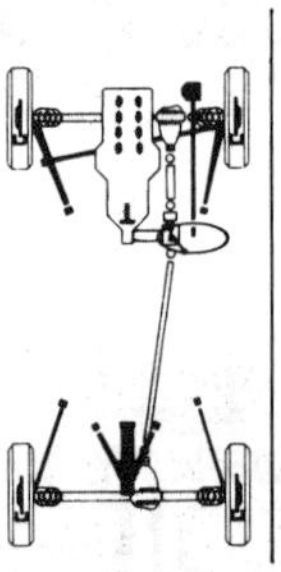

Capacity, cc	3528
Power bhp/rpm	165/4750
Torque lb ft/rpm	207/3200
Max speed, mph	102.4
0-60 mph, sec	12.3
30-50 mph in 4th, sec	5.0
30-70 mph through gears	13.1
mph/1000 rpm	26.5
Overall mpg	16.4
Touring mpg	18.9
Weight cwt	36.8
Boot capacity ft^3	36.2
Payload cwt	14.1

Eighteen years of development have kept the Range Rover on top in the increasingly competitive 4wd market. Offering limousine comfort and formidable cross-country ability, the permanent-4wd Range Rover has gained in performance by adoption of fuel injection on its all-alloy 3528 cc V8 engine. Linked to a ZF 4-speed automatic gearbox, it makes the Range Rover a quick and comfortable motorway cruiser.

FORD SCORPIO 4×4 £19,315

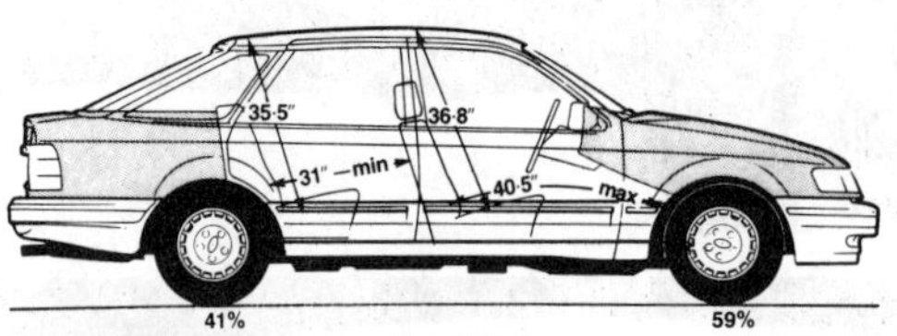

41% 59%

Length 183·8" Width 69·3" Front track 58"
Wheelbase 108·8" Height 54·8" Rear track 58"

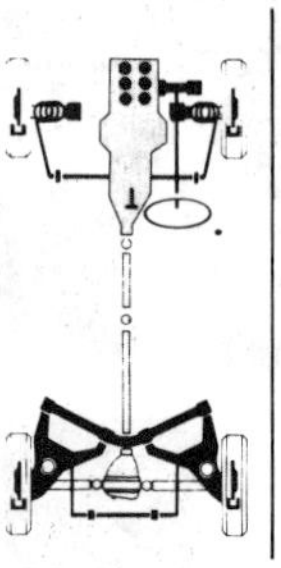

Capacity, cc	2792
Power bhp/rpm	148/5800
Torque lb ft/rpm	162/3000
Max speed, mph	126.4
0-60 mph, sec	9.6
30-50 mph in 4th, sec	8.5
30-70 mph through gears	9.9
mph/1000 rpm	24.1
Overall mpg	18.4
Touring mpg	25.2
Weight cwt	27.9
Boot capacity ft^3	13.8
Payload cwt	9.4

Tested with 2.8-litre engine, Scorpio's chassis is made even better thanks to viscous coupled 4wd with two-thirds/one-third rearward torque bias plus revised suspension settings. Performance adequate rather than strong, economy reflects hard-worked engine and gearchange is long-winded. Incredibly secure handling come rain or shine, with brakes to match. Ride firmer than standard but still excellent, comfort and finish good.

MERCEDES 280 GE AUTO £21,440

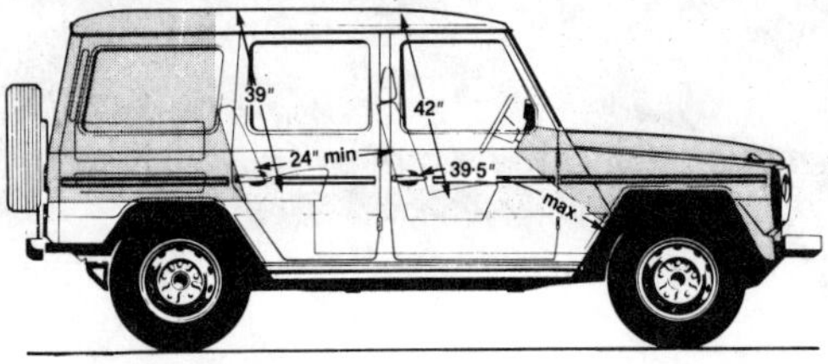

Length 4·39m (173") Width 1·70m (67") Front track 1·42m (56")
Wheelbase 2·85m (112·3") Height 1·97m (77·8") Rear track 1·42m (56")

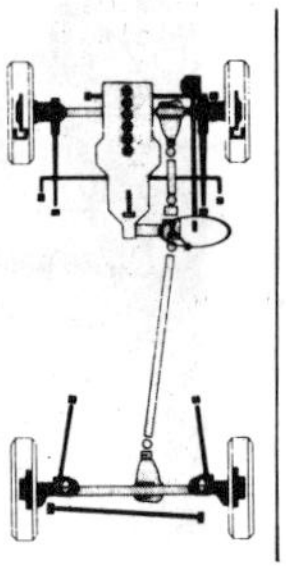

Capacity, cc	2746
Power bhp/rpm	150/5250
Torque lb ft/rpm	162/4280
Max speed, mph	94.4
0-60 mph, sec	14.4
30-50 mph in kickdown, sec	7.4
30-70 mph through gears	19.1
mph/1000 rpm	17.1
Overall mpg	13.8
Touring mpg	16.2
Weight cwt	42.5
Boot capacity ft^3	47.3
Payload cwt	10.9

Mercedes' angular G-Wagen – built in conjunction with Steyr-Daimler-Puch – is designed to tackle the roughest off-road terrain while offering a reasonably lively performance on-road, thanks to its twin cam 2746 cc six-cylinder engine. Its functional interior reflects its main role as an "on demand" rough-terrain vehicle, a task for which its independent differential locks for front and rear axles makes it an excellent performer.

RENAULT ESPACE £13,810

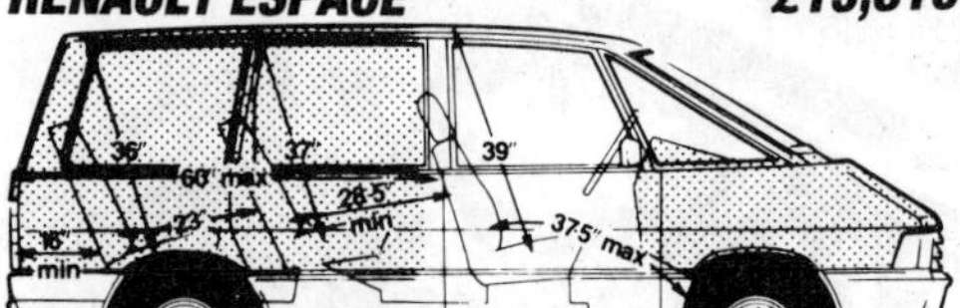

35% 65%

Length 167·3" Width 70" Front track 56"
Wheelbase 101·5" Height 65·3" Rear track 58·5"
Width between w/a 46" Floor height 45·5" max 41" min

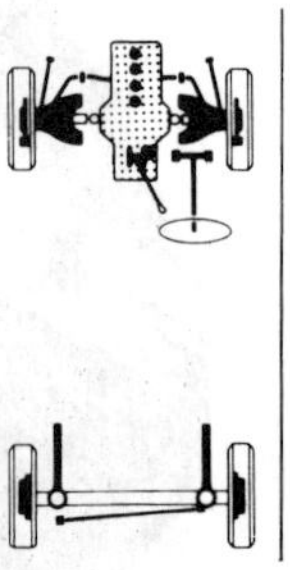

Capacity, cc	1995
Power bhp/rpm	110/5500
Torque lb ft/rpm	120/3000
Max speed, mph	102.8
0-60 mph, sec	11.8
30-50 mph in 4th, sec	11.6
30-70 mph through gears	12.4
mph/1000 rpm	22.8
Overall mpg	21.3
Touring mpg	32.0
Weight cwt	23.7
Boot capacity ft^3	35.3
Payload cwt	11.8

Matra-built Renault Espace is the purist's people carrier. Plastic body panels on a galvanised steel chassis promise longevity, while space age styling looks the part. Driver sits far back and legroom is restricted. All rear seats are easily removed, or convert into table top (rearmost pair is an extra), while the front ones can be swivelled around to create a mobile conference room. Performance, economy and handling are car standard.

VW CARAVELLE SYNCRO £18,447

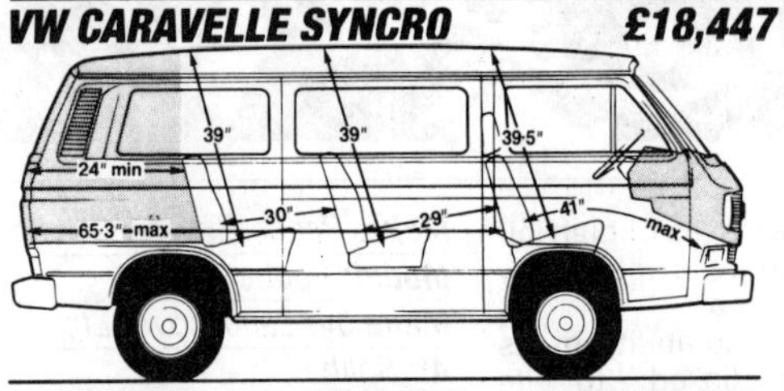

Length 4·57m (180") Width 1·87m (73·6") Front track 1·58m (62·2")
Wheelbase 2·46m (96·8") Height 1·99m (78·3") Rear track 1·57m (61·8")
Width between w/a 60·6" Floor height 43" max 38·4" min

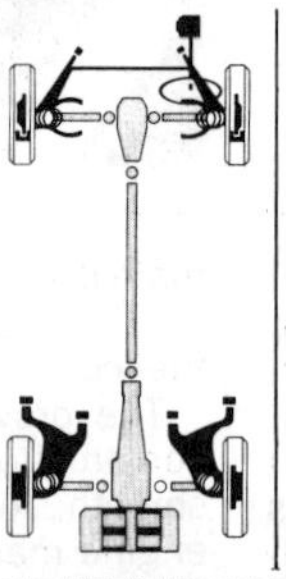

Capacity, cc	2109
Power bhp/rpm	112/4800
Torque lb ft/rpm	128/2800
Max speed, mph	84.5
0-60 mph, sec	16.2
30-50 mph in 4th, sec	16.2
30-70 mph through gears	19.5
mph/1000 rpm	19.9
Overall mpg	20.7
Touring mpg	23.1
Weight cwt	27.8
Boot capacity ft^3	30
Payload cwt	19.9

This luxury 4wd personnel carrier suffers only from its van-like appearance and poor crosswind control. Based on the VW Transporter, power comes from a rear under floor-mounted lusty 2.1-litre fuel injected "boxer" engine, producing 112 bhp, linked to a transmission developed by Steyr-Daimler-Puch – utilising a viscous coupling. Offers armchair comfort for driver and six passengers, good luggage area and off-road capability.

VOLVO 760 TURBO ESTATE, AUTO £20,800

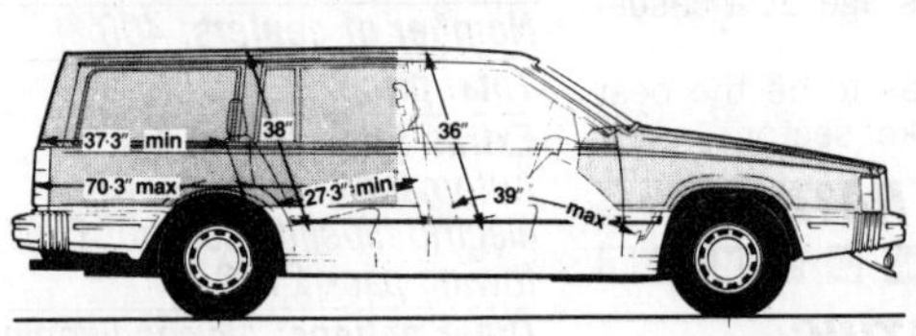

Length 4·79m (188·5") Width 1·76m (69·3") Front track 1·46m (57·5")
Wheelbase 2·77m (109") Height 1·44m (56·8") Rear track 1·46m (57·5")
Width between w/a 43·8" Floor height 32·9" max 29·5" min

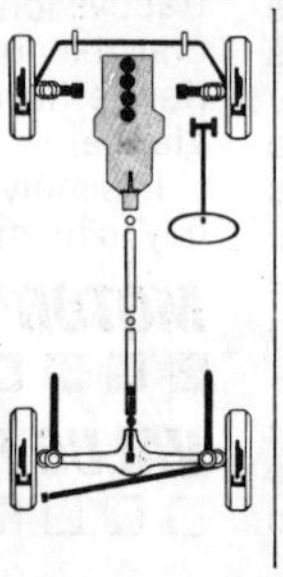

Capacity, cc	2316
Power bhp/rpm	180/5800
Torque lb ft/rpm	192/3400
Max speed, mph	124.0
0-60 mph, sec	7.6
30-50 mph in 4th, sec	7.3
30-70 mph through gears	7.5
mph/1000 rpm	25.2
Overall mpg	20.2
Touring mpg	26.7
Weight cwt	26.3
Boot capacity ft^3	42.4
Payload cwt	9.4

Arguably the first mega-estate, the 760 Turbo combines tremendous performance with enormous load carrying potential. Four-cylinder turbo lacks the smoothness and refinement of rival "sixes", but it is reasonably efficient. More wieldy than popular image suggests, and ride very good for a live axle design. Roomy interior marred by poor driver's legroom. Merits are leather trim, good finish, plenty of equipment and high build quality.

LINE UPON LINE

Land Rover's new flagship, the Range Rover Vogue SE, takes the vehicle one step further away from being a serious mud-plugger. Or does it? Here we take a look at the new model in its natural surroundings and go back to the factory for a glimpse at how a Range Rover is pieced together

● **The current Range Rover production line started work on 16 March 1981**

RANGE ROVER VOGUE SE

Land Rover Limited announced their latest, even more upmarket version of the Range Rover, the Vogue SE, at the Geneva Motor Show earlier this year. Billed as the new luxury 'flagship' of the company's range, the vehicle features as standard air conditioning, an electric tilt-and-slide sunroof, automatic transmission and seats which are fully trimmed in grey leather, the front two being electrically adjustable for tilt, reach and rake. The leather theme is extended to the fascia-mounted grab handle, and both transfer gear lever and handbrake have matching leather gaiters. The luxury touch is also applied to the cappings on the doors and their 'pulls' which are decorated with polished burr walnut. Still more sound-proofing has been added around the transmission tunnel in an effort to quieten the almost traditional, but now much more subdued, whine from the transfer gears. Subdued, in fact, to the point where it is almost a figment of the imagination.

If you are in the market for a Range Rover SE and you are not too keen on Cypress Green or Caspian Blue, then you're out of luck, as they are the only colours in which this model is available. But as a bonus the wheel centres are sprayed to match and you get a solitary coachline thrown in.

Fortunately, underneath this thin veneer of luxurious respectability there lurks the same old Range Rover, ever willing and able to take to the hills and, in the right hands, perform miracles of mobility under trying conditions. The fact that a growing number of customers are apparently willing to pay in excess of £27,000 for the privilege of owning one, largely for on-road use, is, of course, excellent news for Land Rover Limited, and one cannot blame the company for taking advantage of the situation by tailoring the product to suit this market, even if it has resulted in a vehicle that is totally impractical for day-to-day off-roading. One does wonder just how much longer buyers in this bracket will happily go on accepting the still mildly roly-poly cornering, slightly vague handling, transmission noise and tacky plastic fascia in a so-called 'top-of-the-range' model. The Range Rover is undoubtedly a magnificent driving experience, but the novelty does wear off!

However, by way of compensation, the standard ZF four-speed automatic transmission is as silky smooth as ever and the V8 is an absolute delight, despite the fact that it's beginning to get a bit long in the tooth now. It's a funny thing, but we had almost convinced ourselves that the introduction of the Vogue SE would mean an increase in displacement, say 3.9 or 4.0 litres, with power up to at least 200 bhp, to convince customers willing to pay the premium that they are getting value for money. (Did you know 98% of Range Rovers are acquired by private buyers?) We had also rather hoped that the simplistic lockable inter-axle differential in the transfer case would be replaced with a viscous coupling unit. But it was not to be.

It does seem that £5000 or so is a great deal to pay for the doubtful privilege of a sunroof, all-electric front seats, slippery leather trim (just like the Yanks are already getting) and painted wheels. Not even a cruise control! Ah well, perhaps it's the 50th anniversary we should be rooting for!

In all fairness to Land Rover Limited, and despite everything, the Range Rover

LINE UPON LINE

Left: the leather treatment. Now the Americans have had a go it's our turn

Above: trusty old Rover V8

Left: our Lightweight Land Rover stealing the show

is still a magnificent machine. It is streets ahead of the opposition when it comes to all-round ability, but, unfortunately, most owners don't want it for that; for them, 'status' is still the prime factor involved. It's the 'statement' a Range Rover makes about them that is so important. Important enough for over 187,000 sales so far – and annual sales are rising rapidly.

The target for the first year's sales in the United States of America was 3000 vehicles, but Range Rover North America's 60-strong dealer network actually sold 3200. The aim is to increase this figure to a yearly turnover of 4200 units, but that could well be a conservative estimate, judging by the Yanks' reaction to Solihull's baby.

In the growing affluence of the European market place, Range Rover continued its dramatic annual sales increases: up 86% from 5190 in 1983 to 9654 in 1987. In the UK too, things are looking good, with a 68% increase over the same period, rising from 2993 units in 1983 to 5027 in 1987. However, it's not all roses. In the rest of the world (Africa, Middle East, Far East) Range Rover sales are down some 32% from 4006 to 2724, reflecting the general decline in the economies of these areas.

The Land Rover has also suffered some setbacks in most areas, with Europe alone showing a marked improvement, up 44% from 3296 to 4746. Throughout the rest of the world sales are down substantially, once again reflecting the weakening financial positions of these traditional markets. This is a trend also felt by other manufacturers. It is worth noting, however, that in many areas, 20-year-old Land Rovers are undergoing major overhaul locally and being pressed back into service, presumably for another 20 years, so the spares supply situation looks good in the long term. A fact which should delight Land Rover Parts and Equipment (LRPE), and the Spanish 'Santana' side of the operation which is apparently being geared up to cater for this.

There is no doubt that sales of Land Rovers will improve as the moves to provide a more civilised interior progress. However, a great deal more needs to be sorted out before the vehicle is readily acceptable in the market now dominated by the Japanese imports. These already offer car-like interiors and comfort, allied to an off-road performance that is adequate for the recreational aspirations of the majority of their owners.

What is required from the British motor industry, not necessarily Land Rover Limited (though they would be the natural choice), is a new 4×4 vehicle engineered to suit this flourishing sector. Obviously it couldn't be called a Land Rover without harming their worldwide reputation for rugged, durable utility and military vehicles. Perhaps 'Gipsy' could be resurrected, and that isn't intended as a slur on what was a superb four-wheel-drive off-roader in its day. It would also be gratifying to see a British 4×4 estate car along the lines of the impressive Subaru range, perhaps designed with improved clearances to endow it with a more effective and non-damaging off-road ability, rather than out and out all-road performance.

Whether such things will ever come to pass remains to be seen, but the future undoubtedly holds some surprises for us all. Certainly Land Rover staff are being even more tight-lipped than usual about their future plans, and it is difficult to say whether these do include an intermediate vehicle to slot in between Land Rover and Range Rover, as rumour has it. The Range Rover's ride, handling and interior spaciousness could logically be improved by extending the wheelbase to 110". That and more has already been done to individual vehicles, both by the factory and outside companies, with quite impressive results. This would open the way for a new 100" wheelbase machine to replace the Range Rover in its originally intended recreational role, while *it* continues to be developed and marketed as the luxurious, top-of-the-range 'flagship'. We're not saying this will definitely happen, but there is a certain logic in this as a solution to the apparent 'where next?' dilemma.

If you are a current or potential Range Rover/Land Rover owner, how would you like to see them developed?

● Top: The magnificent sight of seven, or is it eight (?), Range Rover Vogue SEs lined up outside Eastnor Castle

'Fortunately, underneath this thin veneer of luxurious respectability there lurks the same old Range Rover, ever willing and ever able to take to the hills'

MAJOR ASSEMBLY

LINE UPON LINE

1 In the beginning; chassis, axles and suspension meet up for the first time

2 Meanwhile, in another part of the factor
production line for initial fitting out

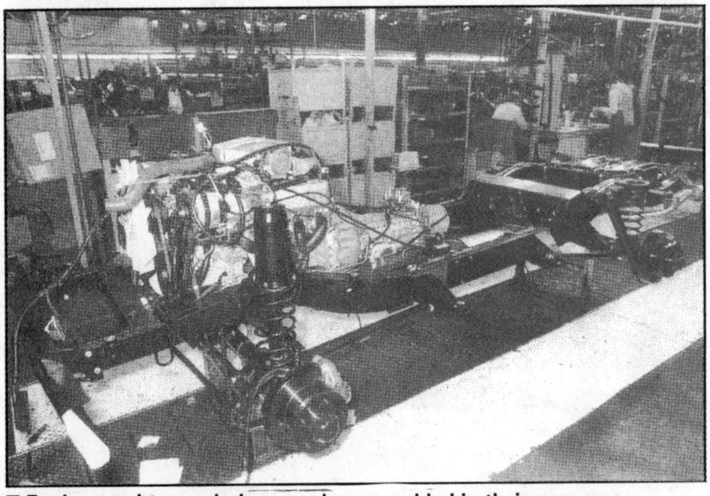

4 Engines and transmission are also assembled in their own area before being passed through, as a complete unit, to join the chassis on the line

5 Partially completed body assembly descends from an overhead trackway to rendezvous with its chassis

8 A quick blast on the air wrench tightens the upper tailgate fixings...

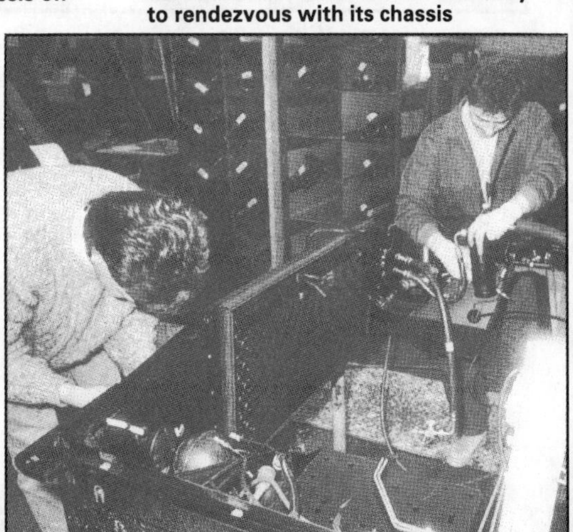

9 ... whilst at the other end, the headlights go into place

ll is lowered into place on the

8 Carefully timed, a roof assembly descends precisely to its final resting place on the bodyshell

6 Remorselessly the line moves on ... and a four-door Range Rover takes shape from the myriad pieces

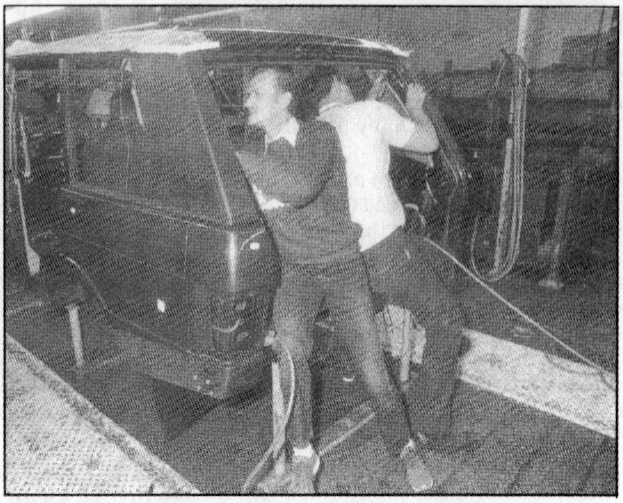

7 The vinyl covered rear quarter panels are quickly secured ... and the line keeps rolling along

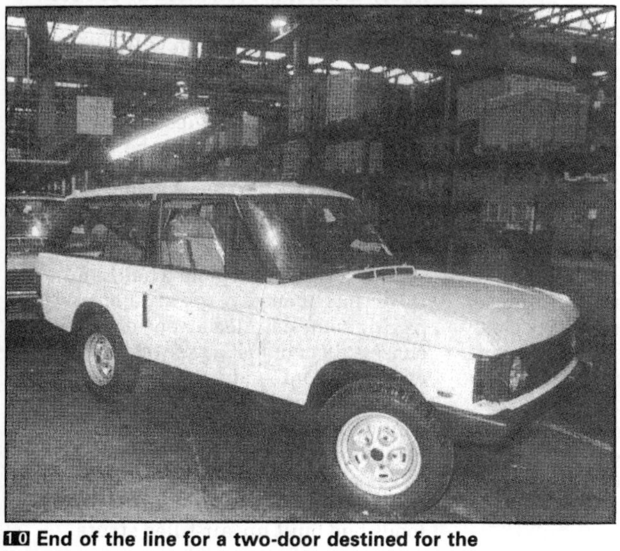

10 End of the line for a two-door destined for the lucrative export market

11 The moment of truth at Quality Control, an essential department in any manufacturing industry

Photographed at Wheal Coates, St Agnes, Cornwall, a property of the National Trust

Little-altered Range Rover exterior hides major changes on Overfinch. Engine is 280bhp Chevy V8 unit

HOME ON THE RANGE

. . . with three special Range Rovers. Report by David Sutherland, photos Paul Debois

The white Range Rover did not attract undue attention as it descended the hill and rolled gently into the north Cornish fishing village. "Holidaymakers," muttered one of the locals into his pint. "Down from London, I expect," observed another, placidly.

Some of the more mechanically-minded local lads looked up to watch the Range Rover squeeze into the square. Not that there was anything so notable about its arrival; simply that watching the driver carefully manoeuvre near on two tons of bulk past boats and fishing nets, only to meet a dead end, was a pastime in itself.

Under the spotlight of the locals' glare, the driver three-pointed, having decided to try another of the small roads that wound around the coastline. He tapped the selector back to D, and gently squeezed the throttle.

From the twin tailpipes there suddenly came a rich, throbbing bellow as the engine began to take the strain of the steep hill back to the top of the cliffs. At that, the local lads began to take a much keener interest; one bumped his head on the bonnet of his Morris Ital as he tried quickly to extract himself from somewhere in the region of the sump.

Not since that video of *Bullit* last week had they heard anything like this! There was sure as hell no Rover V8 under that chunky bonnet — that was rippling American V8 muscle with a carburettor the size of a bucket, and de-smogged with it! Then they took a closer look at the bodywork and saw the tell-tale badges: 'Overfinch' on the leading edge of the bonnet, '570 Sport' on the flanks.

The Range Rover disappeared around the first bend, but it could be heard right to the top of the cliff road and beyond. The woofling V8 lingered in the air; the local lads argued animatedly about how many cubes were in the engine, how many barrels in the carburettor.

The large audible hints being dropped by this Range Rover made it fairly clear in principle what was under the bonnet, if not in detail. The firm responsible was Ascot-based Overfinch Ltd, which has been specialising in engine conversions for Range Rovers for some years now. Before Overfinch came into being, conversions along the same lines went under the name of Schuler.

Calling Overfinch's work mere engine conversion would be an injustice; managing director Douglas Cooke stresses that the Overfinch treatment is much more than that. The entire drivetrain is renewed and the chassis is uprated to give the road-car handling that the standard Range Rover is claimed to have but patently doesn't.

Overfinch offers a number of packages, including a tone-up for the old — rather sluggardly — pre-injection Range Rover. But what we had here was the big daddy of them all; the full engine, transmission, and suspension treatment.

The 570 Sport pack means a new, all-iron Chevy V8 with a four-barrel Rochester carburettor, imported from a main GM dealer in Florida. In standard form it's good for 205bhp at 4000rpm, but alloy heads and high-lift cams from hotter GM cars, plus a new stainless steel free-flow exhaust, bump power up to 280bhp at 5000rpm ("It's actually nearer 300bhp," says Cooke). However, it's the torque that really takes your breath away; the Sport conversion increases the standard GM output from 280lb ft at 1800rpm to 340lb ft at 3000rpm.

As well as throwing away the Range Rover V8, the standard transmission goes, too. Some might be loathe to dump the ZF four-speed, for it's a good 'box, specially adapted to cope with off-road work (a lock-up first as well as top, and special oil pick-ups to cope with surge on steep hills). However, the GM700 'box is also a four-speed with lock-up top, so in effect little is lost.

Replacing the Range Rover's normal transfer 'box is a specially developed Ferguson Formula unit that splits the torque 37 per cent to the front, 63 per cent to the rear (the standard Range Rover distribution is 50/50).

Once you have this lot on board, you're likely to become even more aware of how wobbly the standard Range Rover chassis is; the suspension will give you a luxurious ride over every bad surface known to man, but start to turn corners and it's a joke. Roll angles are less alarming than they used to be, but there's still too much slop through the springs and steering. The brakes aren't going to cope with the Chevy power, either.

So Overfinch has done what Land Rover might one day get around to — paying some detailed attention to the chassis. It's not just a case of firming everything up — any quick-tune merchant can do that — because that would in the process ruin the on and off-road ride.

To provide extra roll stiffness that would not manifest itself until cornering loads had built up, air-filled struts (pressure-adjustable from the dash) are mounted outboard of the coil springs, replacing the single, centrally-mounted self-leveller. The front and rear coils are re-rated, as are the dampers, and the ride height lowered by 1½ inches. To provide good on-road performance, but with also some off-road capability, Overfinch chose Avon Turbosteel 255/60 VRs. Larger front discs are fitted.

The price? Well, all the overtime in the world down at the boatyard ain't going to buy the Cornish lads an Overfinch Range Rover. The 570 Sport conversion costs £18,583, and that includes engine mods, exhaust, brakes, suspension and tyres. So you're talking 19K on top of the standard Range Rover, and that's before you start thinking about cell-phones or neat Italian steering wheels. However, a delivery-mileage Rover V8 EFi engine will return about £1200, according to Cooke. He also says insurance brokers don't go a funny pale colour at the mention of Overfinch; you simply pay an increased excess in the event of a claim.

As far as the Janspeed Range Rover was concerned, man and machine got to know each other on the A303 skirting Salisbury. Not far from where the large Hungarian named Jan Odor has his Janspeed tuning emporium. Now the A303 might be considered the scenic alternative to the M4 and M5 if you're going to the Westcountry, but if you want to get a move on, forget it.

During the summer months lots of steady suburban people hitch up caravans to the back of their cars and drive down this road in too high a gear. They often wave when they see another caravan, and stop for brew-ups in lay-bys in between causing tailbacks that the M25 planners would be proud of. As a kind of joker in the pack in a journey along this road, the dual-carriageways abruptly go back to single-lane roads. Or, just to make things exciting, they become those interesting three-lane roads where the middle lane is the indeterminate no-man's land.

Given a Range Rover that was in a hurry, and a route full of holiday drivers who definitely weren't, it was at least comforting to know that there was something special about it. This Range Rover was fitted with Janspeed's latest turbo kit, developed for the EFi models which appeared three years ago.

Unlike the twin turbos for the old carburettor Range Rovers, the EFi unit is a single turbo, a Roto Master T4 mounted on a high-grade cast-iron manifold, and with boost pressure relieved by Janspeed's in-house designed wastegate. An intercooler and engine oil cooler are fitted (automatic Range Rovers already have a standard transmission oil cooler).

Janspeed Range Rover's power is boosted to 238bhp by Roto Master T4 turbo with intercooler system

The standard Lucas fuel injection is retained, though it operates in conjunction with a Zyteck single-point injection computer. The end result? Power goes up from the standard 165bhp at 4750rpm to 238bhp at 4500rpm. Torque is claimed at 298lb ft at 4000rpm against 207lb ft at 3200rpm.

Janspeed also offers a suspension pack, consisting of 25 per cent firmer front and rear springs. It has also used a rear anti-roll bar to reduce the Range Rover's inherent body roll; not a move everyone would approve of, given that to do this is to risk spoiling the excellent ride quality.

The conversion isn't cheap, but at the same time we're not talking telephone numbers. The turbocharger costs £2410 fitted, and is also listed at £1850 for parts only. The suspension pack is £605.

Hacking down to Cornwall, caravans were picked off by the score; it was pleasing to see them shrinking fast in the rear-view mirror. The extra power and torque provided by the Janspeed turbo boost was partly responsible, but so was the basic design of the Range Rover. You sit high up, and the panoramic vision allows you to see over queues of traffic and hedgerows. Thus, you can safely carry out overtaking manoeuvres that just wouldn't be on in a normal car. The caravanners didn't like that; they flashed their lights and made comments to their passengers about it.

The third in the Range Rover trio was one from the factory, the new Vogue SE, powered by the regular 165bhp V8. This was Solihull's flagship all right, £28,885 worth of Sloane Square ecstasy. Pembroke Grey leather for the seats, gearbox transfer lever and handbrake; polished burr walnut door-cappings; extra sound deadening; lots of equipment, including electrically adjustable front seats, electric sunroof, air-conditioning. Did someone say this used to be an off-road vehicle?

Yes, the Range Rover *used* to be an off-road vehicle, but it's hardly one any more. Are we crazy? Not really, the Range Rover has become a tarmac-lover. Of course it's still the most assured vehicle you can take on to the rough, but when did you last see one doing anything more daring than crossing a muddy car park?

Standard Vogue SE may be tops off the road, but lacks punch for overtaking on the tarmac

No, virtually all the development that has ever been put into the Solihull estate has been to improve its on-road capability; a better, five-speed gearbox, less transmission whine, automatic transmission, electric windows, acres of thick-pile carpeting, and now even *leather* upholstery. Who's going to risk their near-£30,000 luxury vehicle through a quagmire, particularly when they have to detach the front spoiler first? Nowadays you look upon Range Rovers as road cars.

However, recently Land Rover announced some important changes to the transmission, which should benefit on-and-off-road refinement. A new self-locking viscous coupling unit, made by Borg Warner in South Wales, now controls the centre differential, dispensing with the need to lock the centre diff manually. Now, the vc senses the loss of grip and locks up accordingly. The facility to change between high and low ratios remains, of course.

The three Range Rovers met up in Cornwall. A beautiful July evening, with lots of fresh white spray clinging to the rugged coastline; perfect for the 'outdoor' lifestyle led by Range Rover customers.

The driver of the Overfinch had clearly enjoyed himself on the burn down from London, though in doing so had come near to busting his plastic limit at various filling stations. The SE driver eased himself out of the leather

seats and said how warm it was outside without the air-conditioning. The Janspeed driver made a bigoted and unreasonably sweeping statement about caravan owners.

Each of the three had proved impressive in its own way: the Overfinch for its glorious GM V8 conversion; the Janspeed for its ability to deal with the caravans; the SE for its sheer opulence. But gathering them together for comparison highlighted their individual characters.

Apart from the sound effects, the Overfinch is a supremely competent conversion; there is none of the crudeness that a huge American V8 might suggest. The motor becomes slightly wheezy as it touches 5000rpm, but with torque peaking at 2000rpm less it's pointless using the revs anyway.

Perhaps the exhaust is a shade too resonant at mid revs, but the note settles down to a quiet burble at a steady cruising speed.

And the Overfinch certainly moves! It makes the ordinary Range Rover feel distinctly sluggish, and Cooke says that it performs zero to 60mph in just 7.7 seconds, zero to 100mph in 24.5. That's fast enough to embarrass a 16-valve hot-hatch.

But the Overfinch has more to its credit than speed; it also has a subtly improved chassis. As well as checking cornering roll, the suspension changes provide noticeably increased stability at all speeds. Range Rovers, with their high sides, constantly feel as though they're fighting cross-winds, but the Overfinch makes steadier progress. The steering also lacks the normal quarter-turn of vagueness. And no sacrifice in ride comfort either!

The Janspeed is quick all right. At Salisbury they say it will make 60mph in just over nine seconds along with a 125mph maximum (for the standard Range Rover auto, 12.2 seconds and 107mph). On an automatic Range Rover, at any rate, it's a civilised turbo; the boost begins at quite low revs, providing a progressive stream of power. It's quiet, too, apart from the wastegate. It would be interesting to see what the turbo adds to a manual vehicle (manuals still account for a third of sales).

But is it the right thing to turbocharge a Range Rover in the first place? If the standard item has a problem it's the lack of mid-range punch. There's enough low-speed torque to uproot tree-trunks, but overtaking into slim gaps can be a tight-run thing. The Janspeed conversion does give more 3000-4000rpm grunt, but the power is inevitably stacked towards the top end. And you do seem to lose a little of the standard V8's smoothness.

Can the Range Rover Vogue SE really cut it with the BMWs and Jaguars? The autobox — efficient though it is — together with the air-conditioning tax the V8 to the limit; you find more often than not you've run out of throttle to press. It's probably as well that most other drivers don't actually realise how slow a standard Range Rover is — they certainly wouldn't move out of the way so quickly if they did.

The Range Rover is every bit as charismatic as it was 18 years ago. Possibly more so. It has the looks (everyone recognises the lines of a Range Rover), and an unbeatable classiness. It's fun to drive, whether it be through London in the rain or down a rocky Cornish cliff-path. And despite its 18 years, it still has so much development potential — the Overfinch is abundant evidence of that. ■

Viscous coupling transfer box and four-speed automatic provide easier access to exceptional off-road ability. New Michelins give fine road grip without compromising traction on the rough

ROAD TEST

Range Rover Vogue SE

With a new transfer box — the biggest mechanical change since the Range Rover was first introduced — the latest SE enters its 19th year quicker, smoother and quieter than ever

Price *£28,855* **Top Speed** *101mph* **0-60** *11.7secs* **Mpg** *15.8*
For *Quick on tarmac, unequalled off-road* **Against** *Still thirsty*

IF EVER THERE WAS AN ARGUMENT for evolution over revolution, the Range Rover is it. Land Rover isn't in the business of radical change, so over the past 18 years it has honed and refined the Range Rover without ever departing from that instantly recognisable shape.

Clearly it's a policy that works. Sales are running at an all-time high and the changes are still happening. Most are minor, some more fundamental, but overall the Range Rover just gets better all the time. It would be even better with the 3.9-litre V8 just launched in the US (see *Arrivals*), but even without it the latest Vogue SE is quicker, smoother and quieter.

The SE is the most luxurious Range Rover ever sold in the UK. Introduced in April this year, it mates US-spec leather seats with electric adjustment to a standard-fit ZF automatic gearbox, air conditioning, sunroof, a more expensive stereo and burr walnut trim. At £28,855 it's priced well over £5000 above the Vogue. Even without the walnut and leather, the same options would add £3600 to a Vogue.

All 1989 model Range Rovers, from the humblest Turbo D up to the SE, have a new Borg Warner transfer box. Apart from being a good deal quieter, its most significant change is from a mechanical lock for the bevel centre diff to a viscous coupling. It provides progressive and completely automatic locking of the diff if one end of the Range Rover starts to lose traction.

There's also a package of other changes to smarten-up the interior and answer some of the time-honoured criticisms of the Range Rover. New door panels have moulded-in door pulls, recessed door locks and, on the SE, an extra bass speaker in each door. The air conditioning unit has a quieter, more powerful fan and an extra setting to provide warm feet and a cool head.

The stereo has a new aerial in the rear screen, the driver has one-touch window operation and the central locking finally includes the tailgate. The facelift includes heated screen washer jets, variable delay on the wipers and a revised intermittent program for the rear wash/wipe.

Outside, there's only one visible change; the grille now sports the new corporate Land Rover badge. You can also now order an SE in black — although, in line with its price tag, Land Rover calls the colour "Beluga".

PERFORMANCE AND ECONOMY

On paper, the Range Rover's 3.5-litre fuel injected V8 hasn't changed since the Vogue EFi was launched in 1986. The all-alloy pushrod engine still develops 165bhp at 4750rpm and 207lb ft of torque at 3200rpm. But right from the start this particular unit felt smoother through the rev range and more eager than any previous standard Range Rover. What's more, there was little sign of the high rpm pinking present in every previous example we've driven.

Driving through the same ZF four-speed automatic gearbox on the same overall gearing as the last Vogue we tested (6 January 1988), the new SE slices almost a second and a half from its 0-60 time. At 11.7secs it also knocks 0.2secs off the time set by the five-speed EFi. By the time it touches 90mph, the SE leads the old auto by over 10 secs, and is significantly faster in every gear.

It's not clear just where the extra performance comes from. This particular SE scaled 200lb less than the Vogue auto, which could have accounted for some of the difference, although it is a bit of a mystery why it's so much lighter; it can't all be put down to the new ▶

◀ transfer box. Equally, an improvement of this magnitude can't simply have come from fewer power losses in the transfer case. But together with an engine that was plainly put together as its designer intended, this Range Rover proved to be a respectably rapid car.

What hasn't changed is its shape, and with no concessions to low drag bar the under-bumper front spoiler, the Range Rover still runs into an aerodynamic brick wall at 101mph. Manual versions manage an extra 4mph, although the SE might go faster if the ZF box didn't keep kicking down to third. Under full throttle it will kick down even at maximum speed, never in any danger of overspeeding the engine but making sure you either have to run at less than full throttle or well over peak-power rpm to get past 100mph.

Range Rovers don't have tachometer red lines or rev limiters, but 5500rpm is the generally accepted limit for the V8. Flat out through the gears, the ZF box changes up at a fraction over 5000rpm; on part throttle it changes up closer to the engine's 3200rpm torque peak. Up or down, full or part throttle, the changes are about as smooth as you could possibly expect without electronic engine management and transmission control.

Leave the SE in 'D' and the high 26.6mph/1000rpm gearing lets it purr along at 90mph with less than 3500rpm on the clock. The trade off is that low speed response suffers because the ZF box is eager to change up and reluctant to kick down on part throttle. For anything other than motorway work leaving the transmission in '3' is the answer, allowing the injected V8 to make the most of its impressive flexibility. The engine will pull from tickover, with a flood of torque anywhere above 2500rpm. It does get a bit breathless above 5000rpm, but left to its own devices the ZF box does a fine job of keeping the engine away from such heady revs.

Fuel consumption has never been a Range Rover selling point, but the fuel-injected engines are a good deal better than the old carburetted version. Allowing for a 3.5 per cent optimistic milometer, our SE returned an overall 15.8mpg, exactly the same as the last

TECHNICAL FOCUS

For the past 18 years the mechanical layout of the Range Rover has almost been carved in stone. There's still the ex-Buick, ex-Rover all-alloy **3.5-litre V8** *under the bonnet, now fuel-injected and developing 165bhp at 4750rpm with 270lb ft of torque at 3200rpm. In SE form it drives through ZF's familiar four-speed automatic with lock-up in third and fourth.*

New for 1989 is the Borg Warner **transfer box**, *replacing the LT230 unit used since 1982. An epicyclic gearset now provides the high and low ratios, with drive to the centre bevel differential passing through an inverted tooth chain instead of a helical gearset. The mechanical diff lock has gone, replaced by a* **viscous coupling** *that locks progressively and automatically when it detects wheelspin at one output.*

Live axles front and rear provide **permanent four-wheel drive** *through 7ins wide alloy rims and 205R16 Michelin XM+S 244 dual purpose tyres. Standard kit includes power-assisted recirculating ball steering and disc brakes all round.*

A massive box-section steel ladder chassis supports the Range Rover's suspension and drive train. **Coil springs** *front and rear and telescopic dampers provide the necessary wheel movement; leading arms with a Panhard rod at the front and trailing arms with a central 'A' frame at the rear provide the axle location. There are no anti-roll bars but a Boge self-levelling strut at the rear keeps things on an even keel.*

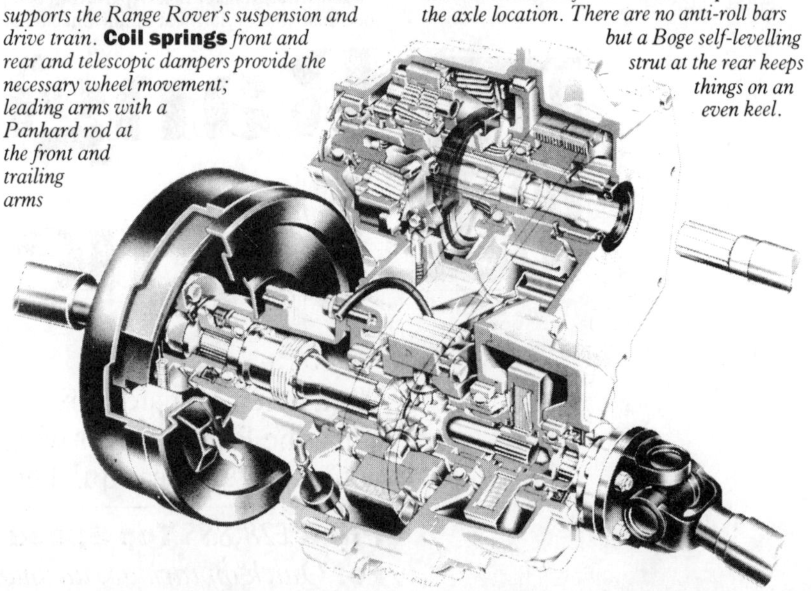

Lack of transfer whine makes on-road cruising quieter than before; at a steady 90mph just a muffled wooffle from under the bonnet and a hint of wind noise intrude. Elevated driving position gives excellent visibility; ZF auto kicks down to third at any speed to provide good overtaking performance

Vogue auto we tested and slightly better than the manual's 15.0mpg. It's around town that the average goes down; on motorways the SE edges the right side of 17mpg. The fuel tank will hold 17.5gals, giving the Range Rover an acceptable touring range of well over 250 miles.

HANDLING AND RIDE

Purely for the road, the Range Rover doesn't give much away to more conventional cars. Land Rover's quite understandable reluctance to fit anti-roll bars means that it rolls substantially, but when you're used to that you can hurl the Range Rover around with considerable verve.

Live axles front and rear mean there are no camber changes despite the long-travel suspension, giving vice-free behaviour. On the limit it breaks into an understeering four-wheel drift, but lifting off the throttle doesn't provoke anything more than a slight tightening of the line.

Michelin XM+S 200 tyres of 205R16 section have been standard Range Rover wear for several years, but 1989 models get the XM+S 244 derivative. According to Michelin, these provide better road performance without compromising off-road ability. On the SE, they certainly generate plenty of lateral grip and also prove tenacious in the wet.

Stability at speed is a lot better than earlier Range Rovers, but the power-assisted recirculating ball steering is slow to respond and feels slightly dead around the straight-ahead. That's partly due to the damper fitted to tame kick-back off the road, and on tarmac it takes a little getting used to. The level of power assistance is just right to retain some measure of feel, and at 3.5 turns lock to lock gearing is acceptably high.

At around two tonnes — 4000lb — there's a lot of Range Rover to stop, so the brakes need to be good. The 11.8ins front and 11.4ins rear discs are up to the job, and although the pedal action is initially slightly spongey the brakes always feel powerful and reassuring. After repeated high-speed stops they fade slightly and progressively, but soon recover fully.

Despite the unsprung weight of two live axles, the Range Rover's long-travel coil spring suspension delivers an exceptional ride. Always supple and well-controlled, it easily soaks up badly patched roadworks and you hardly notice unexpected speed bumps: the rougher the road, the better the suspension works. On very smooth surfaces it can't quite match the magic carpet qualities of a Jaguar XJ6, but it's a good deal better than most German luxury saloons.

The best news is that all the extra road ability that comes from 18 years of development hasn't compromised the Range Rover's essential off-road qualities one iota. For serious work you should follow Land Rover's advice and remove the front spoiler — although this is a fiddly job that could be made much easier with some thoughtful design — but drive into a field full of mud and you don't even need to stop.

The combination of a viscous coupling centre diff lock and the auto box means off-roading a Range Rover is easier for a novice than ever before. There's no need to think about whether or not to engage the diff lock, and the progressive locking effect has distinct benefits if traction is marginal.

Descending slippery slopes, for example, the coupling makes sure that the rear wheels don't lock first and the Range Rover can come down without needing to have any throttle fed through the drive train. On the road, the coupling can react almost instantly to preserve traction on icy surfaces.

Changing from high to low range and vice-versa is a much slicker and quieter affair than it used to be with the old transfer box. Low range first gear with the ZF gearbox isn't as low as on the manual Range Rover, so ultimately there is less engine braking available on very steep slopes. That's offset by the auto's ability to cope with climbs — just leave it in D and the box selects exactly the right gear and can change if necessary half way up the hill without breaking traction.

Without anti-roll bars and with its long travel coil springs, the Range Rover has exceptional axle articulation. That means it can keep all its wheels on the ground where other off-roaders have a pair of wheels pawing the air — and with the fine grip of the Michelin XM+S tyres the Range Rover has traction available without needing the diff locks necessary on something like a Mercedes G-Wagen.

AT THE WHEEL

In the Vogue SE, the Range Rover's usual huge armchairs are upholstered in top-quality hide. The seats were never over-supportive laterally, and the smooth leather means it's difficult to hold yourself in place when cornering hard. Apart from that, the seats are well-shaped and extremely comfortable for long periods.

Electric adjustment via a small control panel on the inner side of the seat provides slide, backrest angle and independent height control of the front and rear of the base cushion. A fold-down armrest is also ajustable for angle. A problem with the base Vogue is that the optional electric tilt-slide sunroof reduces headroom by well over an inch. That sunroof is standard in the SE; but using the seat base height control can restore sufficient room.

There's plenty of seat travel for even the tallest driver, but a non-adjustable steering column means the wheel is a bit of a stretch with the seat towards the rear of its travel. Pedals are large and well-positioned, with a left footrest fixed to the side of the transmission tunnel. A moulded plastic gearshift is a bit out of character with the rest of the interior, and the detent release button that locks you out of neutral, first, second and reverse can be awkward for unfamiliar drivers.

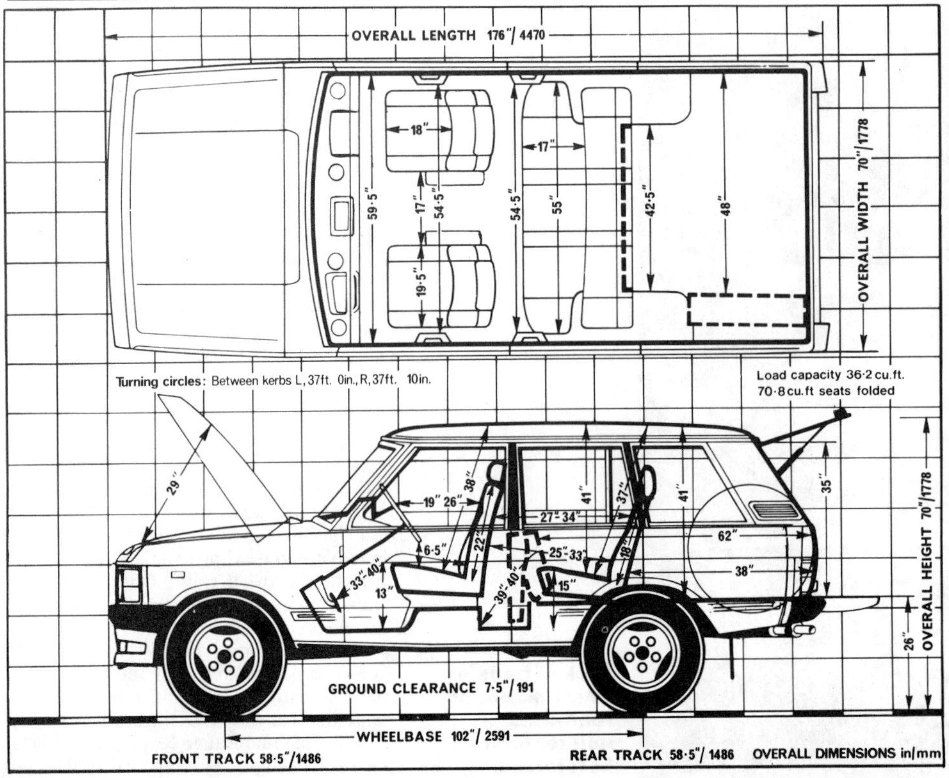

RANGE ROVER VOGUE SE

SPECIFICATION

ENGINE
Longways, front, permanent 4-wheel drive. Head/block al alloy/al alloy. 8 cylinders in 90deg V.
Bore 88.9mm, **stroke** 71.1mm, **capacity** 3528cc.
Valve gear ohv, 2 valves per cylinder.
Compression ratio 9.35 to 1.
Ignition and fuel system electronic ignition, Lucas 'L' fuel injection.
Max power 165bhp (PS-DIN) (123kW ISO) at 4750rpm. **Max torque** 207lb ft (281 Nm) at 3200rpm.

TRANSMISSION
4-speed automatic.

Gear	Ratio	mph/1000rpm
Top	0.728	26.6
3rd	1.000	19.3
2nd	1.480	13.1
1st	2.480	7.8

Final drive ratio 3.54. Transfer box, high ratio 1.222, low ratio 3.321. Bevel centre differential with viscous coupling lock.

SUSPENSION
Front, live axle, radius arms, Panhard rod, coil springs, telescopic dampers.
Rear, live axle, radius arms, 'A' frame, coil springs, telescopic dampers, self levelling.

STEERING
Recirculating ball, power assistance. 3.5 turns lock to lock.

BRAKES
Front 11.8ins (298mm) dia discs. **Rear** 11.4ins (290mm) dia discs. Vacuum servo.

WHEELS/TYRES
Al alloy, 7ins rims. 205 R16 Michelin XM+S 244 tyres.

COSTS

Prices

Total (in GB)	£28,855.00
Road tax, delivery, no. plates	£525.00
Total on the road	£29,380.00
Insurance group	7

SERVICE & PARTS

	Interval (miles)		
Change	**6,000**	**12,000**	**24,000**
Engine oil	Yes	Yes	Yes
Oil filter	Yes	Yes	Yes
Gearbox oil	No	No	Yes
Spark plugs	No	Yes	Yes
Air cleaner	No	Yes	Yes
Total cost	£90.04	£181.58	£252.94

(Assuming labour at £33.35 per hour inc VAT)

PARTS COST (inc VAT)

Brake pads (2 wheels) front	£50.01
Brake pads (2 wheels) rear	£56.32
Exhaust complete	£372.23
Tyre — each (typical)	£94.86
Windscreen	£255.08
Headlamp unit	£30.97
Front wing	£164.56
Rear bumper	£97.08

WARRANTY
12 months/unlimited mileage

Produced and sold in UK by
Land Rover Ltd
Lode Lane
Solihull,
W. Midlands
Tel: 021-722 2424

EQUIPMENT

Automatic	●
Power steering	●
Steering rake adjustment	N/A
Steering reach adjustment	N/A
Head restraints F/R	●/●
Seat back recline	●
Seat cushion tilt	●
Seat height adjustment	●
Split rear seats	●
Heated seats	N/A
Door mirror remote control RH/LH	●/●
Electric windows F/R	●/●
Heated rear window	●
Sunroof	●
Headlamp wash	●
Tailgate wash/wipe	●
Tinted glass	●
Auxiliary driving lamps	●
Central locking	●
Luggage cover	●
Metallic paint	●
Radio/cassette	●

● Standard N/A not available

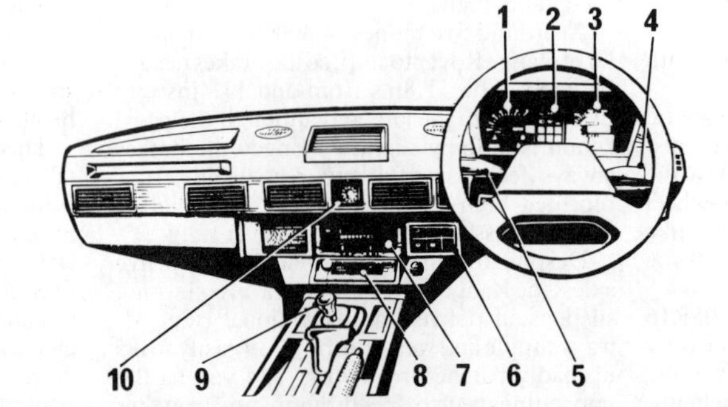

1 rev counter **2** Warning lights **3** Speedometer **4** Wash-wipe **5** Indicators **6** Auxilliary switch panel **7** Air conditioning **8** Stereo **9** Transfer lever **10** Clock

PERFORMANCE

MAXIMUM SPEEDS

Gear	mph	km/h	rpm
Top (Mean)	101	163	3800
(Best)	102	164	3850
3rd (Mean)	101	163	5250
(Best)	102	164	5300
2nd	72	116	5500
1st	43	69	5500

Standing ¼-mile: 18.9secs, 73mph
Standing km: 35.0secs, 91mph

ACCELERATION FROM REST

True mph	Time (secs)	Speedo mph
30	4.2	34
40	5.9	45
50	8.4	56
60	11.7	66
70	16.4	77
80	21.4	87
90	29.5	98
100	43.7	108

ACCELERATION IN EACH GEAR

mph	Top	3rd	2nd	1st
10-30	—	—	—	3.2
20-40	—	—	—	3.6
30-50	—	—	5.4	
40-60	—	—	6.3	—
50-70	—	—	8.1	—
60-80	—	11.1	—	—
70-90	—	12.6	—	—
80-100	—	22.1	—	—

FUEL CONSUMPTION
Overall mpg: 15.8 (17.9 litres/100km)
Touring mpg*: 17.9mpg (15.7 litres/100km)
Govt tests mpg: 14.6 (urban)
26.2 (56mph)
20.2 (75mph)
Grade of fuel: premium 4-star (97 RM)
Fuel tank: 17.5 Imp galls (80 litres)
Max range*: 313 miles
Mileage recorder: 3.5 per cent long
* Based on Government fuel economy figures: 50 per cent of urban cycle, 25 per cent each of 56/75mpg consumptions.

BRAKING
Fade (from 73mph in neutral)
Pedal load for 0.5g stops in lb

	start/end		start/end
1	15-20	**6**	25-45
2	20-20	**7**	25-45
3	25-20	**8**	25-50
4	20-30	**9**	25-50
5	25-35	**10**	25-55

Response (from 30mph in neutral)

Load	g	Distance
20lb	0.34	88ft
40lb	0.65	48ft
60lb	0.85	35ft
65lb	0.90	33ft

WEIGHT
Kerb 4192lb/1903kg
Distribution % F/R 49/51
Test 4555lb/2068kg
Max payload 1337lb/607kg
Max towing weight 8811lb/4000kg

TEST CONDITIONS

Wind	3-5mph
Temperature	10deg C (50deg F)
Barometer	1016mbar
Humidity	40 per cent
Surface	dry asphalt/concrete
Test distance	855 miles

Figures taken at 1900 miles by our own staff at the Lotus Group proving ground at Millbrook.

New grille badge only clue to changes under skin. Vogue SE gets leather seats, electric adjustment, walnut trim. New door panels have moulded pulls, recessed locks. Steering wheel offset to show instruments. All alloy 3.5-litre fuel-injected V8 is smoother than ever. Impressive rear room, but split-fold seats shaped for two. Weight-bearing tailgate extends load length

The steering wheel is a chunky rimmed and leather covered affair that's good to grip. It is offset upwards slightly to give a good view of the instruments, which means it can feel slightly strange when you're feeding the rim back through your hands. Two main stalks comfortably within reach look after the major functions: horn, dip and indicators on the left and two-speed plus variable intermittent wash-wipe on the right. Smaller switches on the steering column turn on the lights and rear wash/wipe.

Instruments are basic but clear and easy to read. A rev counter and speedo flank gauges for fuel and temperature, with a clock in the centre of the dash panel. A switch panel to the lower left of the steering column looks after rear fog, heated rear screen, driving lamps and the single interior lamp but it's easy to confuse them in the dark.

Standard-fit air conditioning does a good job of keeping the car cool, but lacks automatic temperature control. It's a little quieter than before, and boasts a new setting that can provide chilled air through the face level vents while directing heated air through the screen demist and footwell vents — useful in summer rainstorms.

With such a high driving position and acres of glass, visibility from a Range Rover is excellent. You can see ahead over the top of most cars — gaining an edge in safety — and the only slight blindspot is provided by the thick rear pillars.

Pushed hard Vogue SE rolls but grips well; seats lack lateral support. On the limit, Range Rover pushes into an understeering four-wheel drift, but lifting-off brings it smoothly back on line. On slippery surfaces viscous coupling locks centre diff progressively to preserve traction. For serious off-road work, front spoiler unbolts

COMFORT AND SPACE

Cruise at an indicated 90mph and the Range Rover cossets you like a top-notch luxury saloon. There's a muffled woofle from under the bonnet, no detectable tyre or road noise and only just enough wind noise from the front pillar to make its presence felt.

The surprise for anyone who's familiar with Range Rovers will be the complete silence from the transfer box. Gone is the ever-present whine and much of the drive-line shunt, replaced by only a hint of gear noise on the over-run. It makes the biggest-ever single improvement to the Range Rover's levels of refinement.

Range Rovers lose the services of the glove compartment when an air-conditioning unit is fitted, but make up for it with the large lidded box between the front seats. Folded back, the lid makes a handy drinks table. There are two small trays on top of the dash, but the new trim panels means thinner pockets in each door.

Even with the seats well back there's plenty of rear leg room, although the rear seat is moulded to provide seating for two, not three. There's more than enough shoulder room, but the centre passenger just isn't going to be as comfortable as the other two.

Behind the rear seats there's 36.2cu ft of luggage space, although the spare wheel takes up a good chunk of one side. Asymmetric folding of the rear seats gives the Range Rover hatchback versatility, and with both folded there's 70.8cu ft available. Fold the lower half of the tailgate flat and it forms a full load-carrying extension to the load floor.

FINISH AND EQUIPMENT

As well as leather upholstery, the SE gets polished walnut cappings on dash, doors and the centre console. The new door panels actually have less wood than before, but with moulded-in door pulls, recessed locking buttons and carpet panels they integrate better with the rest of the interior.

All the upholstery is beautifully stitched, the trim panels perfectly aligned, and there's an air of integrity to the interior that's never quite been there before.

That impression is reinforced when you drive hard over poorly surfaced roads; there isn't a single squeak or rattle.

With a body that's been in production for such a long time, panel gaps on the range Rover are uncomfortably wide by current standards. Alloy panels mean that it's easy to spot slight ripples along the sides, but even in black (or rather "Beluga") the paint finish is beyond criticism.

Central locking to the tailgate is a long-overdue upgrade, and the heated screen washer nozzles are a valuable winter aid. The six-speaker stereo benefits from the un-damageable rear screen aerial; FM reception is as good as with most external aerials.

However, despite the extra pair of door speakers now fitted to the SE, sound quality is way down on what you've a right to expect in a £29,000 car; to take advantage of the new found silence from the transfer box, a drastic improvement is still required.

There's not a great deal more you could ask for in the way of standard equipment: central locking, electric windows and mirrors, air-conditioning, and the sunroof are all included. The only listed optional extras for the SE are front mudflaps, heavy-duty front suspension and a towing pack.

VERDICT

There's a lot more to the new Range Rover than a quieter transfer box. Combined with the SE's automatic gearbox it makes taking advantage of its exceptional off-road ability almost child's play. And it looks as if the build quality problems that affected the Range Rover in the past have been laid to rest. Certainly, our test SE was smoother, quieter, faster and better screwed together than any Range Rover we've tested before.

Apart from a couple of extra doors and a spoiler, the Range Rover looks just as it did at its debut 18 years ago. Be in no doubt, though, that today's Range Rover is better than ever. There isn't another car in the world that could stay on top with so few changes, but the Range Rover just gets better and better. The latest Vogue SE is even further ahead of the pack than before. ■

VW CARAVELLE SYNCRO 2.1 — £19,764

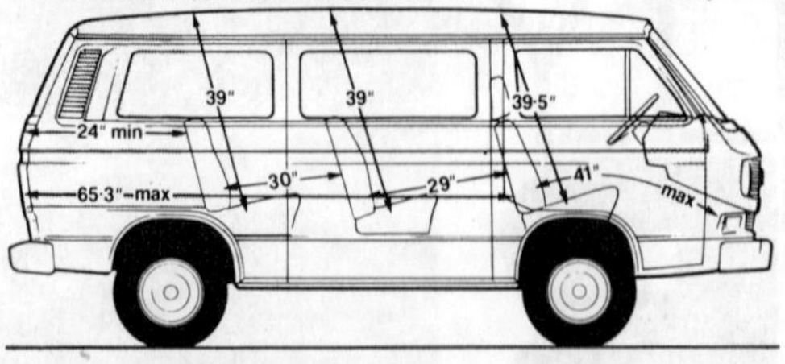

Length 179.9" Width 72.6" Front track 61.7"
Wheelbase 96.5" Height 78.3" Rear track 61.4"

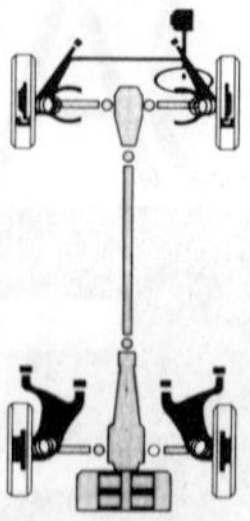

With drive through a viscous coupling and diff locks front and rear, only a lack of ground clearance can stop the Caravelle syncro. Van-like shape limits top speed, but if you want a luxury minibus that can carry seven people and their luggage in comfort whatever the conditions, then VW's four wheel-drive people carrier has few competitors.

Capacity	2109cc
Power	112bhp at 4800rpm
Torque	128lb ft at 2800rpm
Max speed	92mph
0-60mph	17.2secs
30-50 in fourth	15.3secs
30-70 through gears	24.5secs
ss ¼ mile	20.6secs
MPG overall/touring	19.2/21.0
MPH/1000 revs in top	17.8mph
Kerb weight	3661lb
Drag coefficient	N/A
Tested	6.8.86

MERCEDES 280GE — £23,590

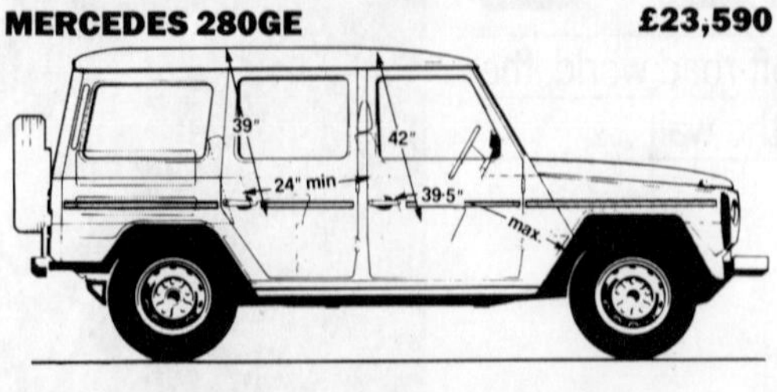

Length 173.0" Width 67.0" Front track 56.0"
Wheelbase 95.0" Height 79.0" Rear track 56.0"

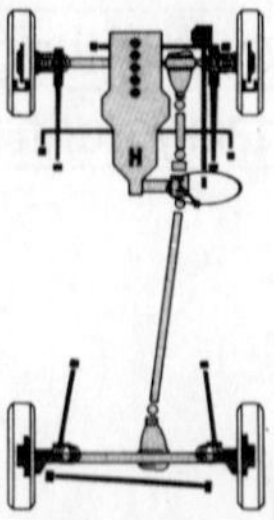

A high price but rather utilitarian looks mean the G-Wagen has never been popular in the UK despite its ability. With the optional four-speed auto box the 2.8 litre petrol engine provides reasonable performance and a good spread of torque. Part-time four-wheel drive with front and rear diff locks provide good traction, coil springs a good ride.

Capacity	2746cc
Power	154bhp at 5250rpm
Torque	167lb ft at 4250rpm
Max speed	92mph
0-60mph	14.4secs
30-50 in fourth	23.3secs
30-70 through gears	15.7secs
ss ¼ mile	19.5secs
MPG overall/touring	12.7/13.9
MPH/1000 revs in top	21.8mph
Kerb weight	4837lb
Drag coefficient	N/A
Tested	6.1.88

LAND ROVER ONE TEN COUNTY V8 — £14,430

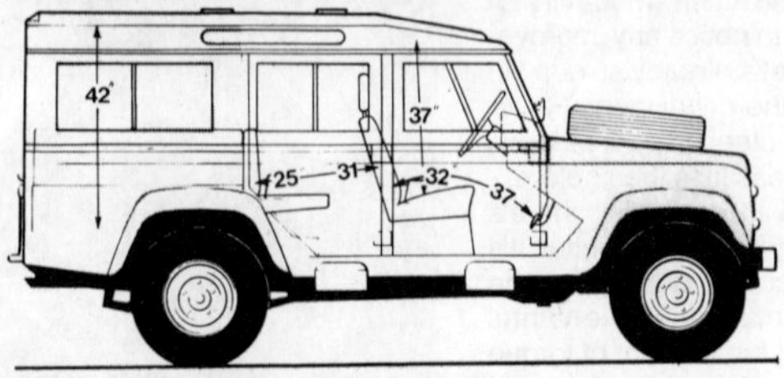

Length 181.0" Width 70.5" Front track 58.5"
Wheelbase 110.0" Height 80.1" Rear track 58.5"

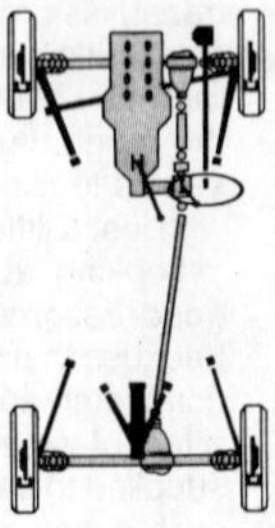

Nominally a twelve-seater for tax reasons, but you'd be pushed to get a dozen adults in the One Ten. Flexibility of carburetted V8 makes the County a willing performer with effortless off-road ability, but huge frontal area restricts motorway ability. Similar drive train, much cheaper than a Range Rover but lacks creature comforts and performance.

Capacity	3528cc
Power	134bhp at 5000rpm
Torque	187lb ft at 2500rpm
Max speed	87mph
0-60mph	15.1secs
30-50 in fourth	11.1secs
30-70 through gears	18,3secs
ss ¼ mile	19.8secs
MPG overall/touring	13.4/14.7
MPH/1000 revs in top	23.0mph
Kerb weight	4404lb
Drag coefficient	N/A
Tested	9.12.87

AUDI 200 QUATTRO AVANT — £31,992

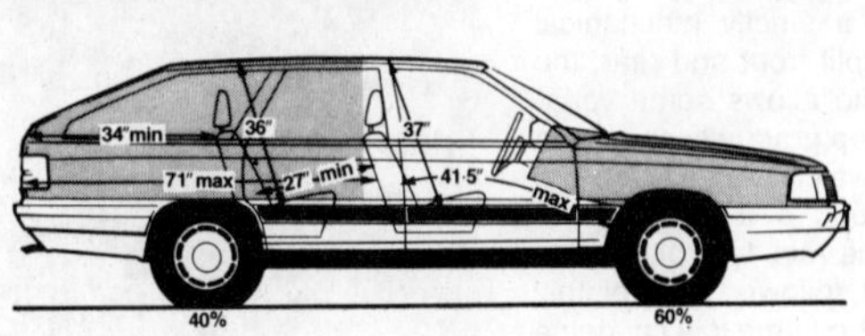

Length 188.7" Width 71.4" Front track 57.8"
Wheelbase 105.8" Height 56.0" Rear track 57.7"

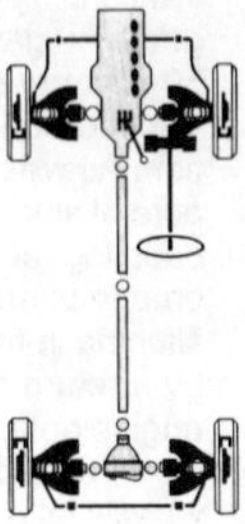

The most expensive Audi you can buy, the 200 Avant quattro combines Audi's expertise in four-wheel-drive with the turbocharged five-cylinder engine and 200 estate body. Superb high speed load carrier, excellent stability and fine performance plus the all-weather security of permanent 4wd and anti-lock brakes make the car almost unique.

Capacity	2144cc
Power	182bhp at 5700rpm
Torque	186lb ft at 3600rpm
Max speed	139mph
0-60mph	7.3secs
30-50 in fourth	7.1secs
30-70 through gears	7.6secs
ss ¼ mile	15.3secs
MPG overall/touring	19.0/20.9
MPH/1000 revs in top	23.36mph
Kerb weight	3387lb
Drag coefficient	0.35
Tested	27.11.85

MITSUBISHI SHOGUN TD — £16,625

Length 181.0" Width 66.0" Front track 55.0"
Wheelbase 106.0" Height 74.0" Rear track 54.0"

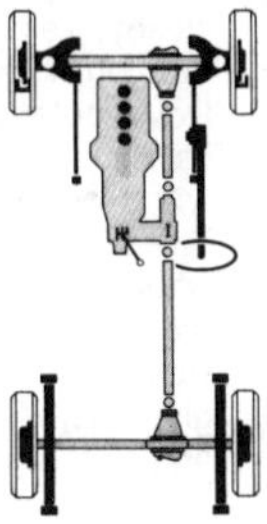

The Shogun's 2.5-litre four is one of the smoothest and most flexible turbo diesels around. Coupled to a delightful five-speed gearbox it provides lively performance, with a high level of refinement and good fuel economy. Part time four-wheel-drive and independent front suspension provide good off-road performance, rear leaf springs biggest drawback.

Capacity	2477cc
Power	84bhp at 4200rpm
Torque	148lb ft at 2000rpm
Max speed	83mph
0-60mph	17.8secs
30-50 in fourth	8.3secs
30-70 through gears	22.8secs
ss ¼ mile	20.6secs
MPG overall/touring	22.1/24.3
MPH/1000 revs in top	20.1mph
Kerb weight	3872lb
Drag coefficient	N/A
Tested	6.5.87

RANGE ROVER VOGUE SE — £28,855

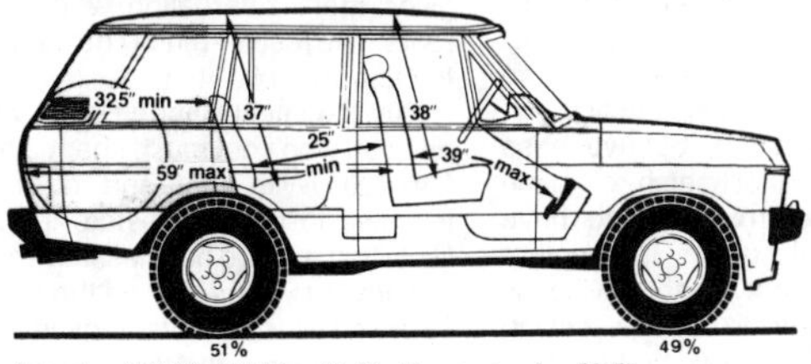

Length 176.0" Width 70.0" Front track 58.5"
Wheelbase 102.0" Height 70.0" Rear track 58.5"

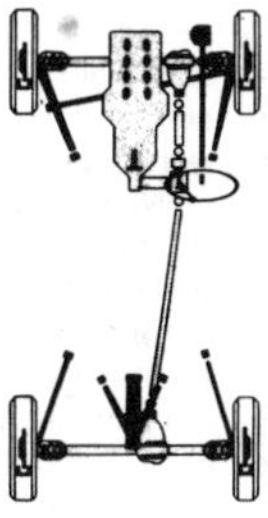

New Borg-Warner transfer box makes Range Rover much quieter, while viscous coupling locks centre diff automatically to preserve traction. Smooth ZF auto box makes off-roading easy, well matched to V8 torque on tarmac. Surprising cornering grip and performance on the road plus unmatched ability on the rough are still tempered by heavy thirst.

Capacity	3528cc
Power	165bhp at 4750rpm
Torque	207lb ft at 3200rpm
Max speed	101mph
0-60mph	11.7secs
30-50 in fourth	N/A
30-70 through gears	12.2secs
ss ¼ mile	18.9secs
MPG overall/touring	15.8/17.9
MPH/1000 revs in top	26.6mph
Kerb weight	4192lb
Drag coefficient	N/A
Tested	23.9.88

AUTOMATICALLY BETTER

An automatic diff lock has been a long time coming for the doyen of the off-road world, the Range Rover. Driving the new model proves it was certainly worth the wait . . .

No longer is the time-honoured tussle with the diff lock lever a necessity for Range Rover drivers. The new system, although a little late in appearing, is simplicity itself in operation

We missed not having to engage the central diff lock on the Range Rover. After more than a dozen years of driving new versions of this Best of British all-terrain vehicle it had become almost instinctive to pause just before leaving the tarmac to lean forwards and latch that stubby transfer case lever across to the left.

But then we missed all the shunting and clonking as well, that inherent jerkiness in the transmission as the Range Rover lurched up steep gradients and dived down into seemingly bottomless pits, the one thing that always suggested that the car was nothing more than a tractor in a limousine's clothing.

And we missed the whine as well, that howl from the transfer gears that just managed to take the edge off the Range Rover's highway cruising refinement. No more of that now, just the smooth, steady pull that you'd really expect from such an expensive luxury machine.

And about time, too. The argument has always been that inherent strength in the transmission of this most genuine of true off-road cars must be more important than ultimate refinement, just in case it ever does get used in really rough conditions, and has to be driven for extended periods in low ratio gears. Good, strong, solid gears in the transfer case were the traditional answer, and they certainly served the Range Rover well for nearly a decade and a half.

As the price of the car soared ever higher, though, launching the Range Rover out of the mud and into the country club car park – and into the heat of an increasingly fierce battle for sales in the four-wheel-drive luxury estate market – it became obvious that something had to be done.

There's been an answer around for years, of course, a system FF Developments has been advocating for all sorts of cars, a mechanism that has been adopted by Ford and BMW in their four-wheel-drive saloons. It uses a big, heavy-duty chain to take drive from the main shaft to the forward propeller shaft, tautly tensioned to take all the slack – and therefore the shunt – out of the system, and at the same time eliminating the whine set up by gear-tooth-to-geartooth contact.

There's even an added advantage. A transfer case in a car with permanent four-wheel drive has to include a differential gearset to prevent unwanted tensions in the transmission when driving on firm surfaces. In the rough, though, that differential has to be locked up or the Range Rover would be at a disadvantage against off-road cars with selectable four-wheel drive. The new chain-drive transfer box solves that problem by employing the technological wizardry of the viscous coupling, which effectively locks up the differential the instant any wheel begins to lose traction and spin. And, because it works completely automatically, there's no need for a diff lock lever in the driving compartment.

Still, long experience of the car made us hesitate just before charging into the rough; a quick lean forwards to reassure ourselves that we really didn't need to lock that diff; no, the transfer lever, though in the same position, now has only high and low range positions.

We want low range, anyway, so we prepare for the usual battle, creeping the car gently forwards and preparing to lean hard on the lever to slot it through its baulky gate into mesh just at the right moment. Surprise! The lever slots forwards with a knife-through-butter feel. That improvement alone could almost justify the switch to the new transfer box.

With that lusty 3.5-litre V8 in front the Range Rover has always been a very relaxing car to drive in the rough. Point the Range Rover at seemingly impassable obstacles and with little more than a disdainfully throaty grumble it will scale incredible slopes, lope languidly across the most tortuous terrain or descend scornfully through swirling streams. Only now it seems to do it even more easily. Is it just because the transmission is so much smoother? Of course it's hard to notice any improvement in a car that's already so good in the rough, but the new Range Rover seems to climb steep slippery slopes with just a little less fuss than before.

Looking at it theoretically, there's good reason why it should be better. It's because that viscous coupling has the completely automatic and quite natural effect of varying the amount of torque supplied to each of the driving axles in such a way as to give greater driving force to the wheels with the greater traction. The old gear-only transfer gearbox gave a strictly mechanical 50-50 torque split front and rear; the viscous coupling allows some variation. As we creep gradually across an acre of sludge we can actually feel the coupling at work; a momentary increase in engine revs tells of a wheel slipping in front, followed immediately by a return to the same low chugging engine note and a slight change in the car's attitude as the grippier rear wheels take up more drive.

We come to a sudden, steep drop, damp and liberally patched with slushy, slimy mud. We've been through this sort of terrain with a Range Rover before. Into first gear and let the car pick its own way down, but ready with a foot poised over the accelerator. The accelerator? Certainly. The front wheels take all the strain in this manoeuvre, with the bulk of the car's weight pressing down on them. That gives them a better chance of digging in and gripping, but it also means that if they start to slip on a muddy patch they could easily lock up and skid all the way to the bottom in an ungainly and possibly quite dangerous slide. To return grip to a sliding front wheel in this situation you have to accelerate, to return rotation and regain traction.

We're ready for it, but we ease through patch after patch of slime without losing control. It shouldn't have surprised us, of course. It's another advantage of that 'magic' viscous coupling's ability to switch driving torque about. Under engine braking it actually reduces the torque reaction from the heavily laden front wheels an gives more retardation work to th rears. The result – much finer all-whee traction control.

Buy the latest Range Rover Vogu and you won't get much change out c £30,000. With that sort of mone invested in a luxury estate it's qui likely that a proud businessman owne will never take the car off the tarma There's still a big advantage in pure on-road driving, though, because th lack of shunt and transmission whin

shows up on the highway as well as in the mud. That transmission howl used to be the biggest source of noise in the Range Rover when driving at motorway speeds, and efforts in past years to refine the car's suspension to allow 100mph-plus performance have not, until now, been matched by an equivalent improvement in noise refinement.

Land Rover has traditionally kept transmission engineering in-house, partly to ensure adequate quality for the workhorse machines – naturally the Range Rover transfer case was the same unit used in Land Rovers – and partly to keep expenditure down.

The need for better refinement in the luxury model has eventually persuaded the firm to go to one of the world's leading gearbox specialists. The new transfer box is made by Borg Warner, a firm not usually recognised as a manufacturer of four-wheel-drive transmissions, being better known in Europe for supplying automatic gearboxes and five-speed transmissions for a wide range of mass-market cars. However, Borg Warner production of 4x4 transmissions in the USA is approaching half a million units a year.

The transmission for the Range Rover is built by Borg Warner at its plant at Kenfig, South Wales, using the basic components – a planetary-type gearset and a Morse Hy-Vo chain – that have proved successful in the US linked to big-torque V8s, but housed in a case that is virtually a 'bolt-on' replacement for the Land Rover unit, fitting directly to the adaptor housings used behind the Range Rover's manual or automatic gearbox.

Now, and then

There can be no other vehicle in the world that combines the inside space of a small van with the performance and refinement of an executive luxury saloon, the nimbleness of a mountain goat and a Herculean towing capacity all wrapped in a package which has won praise in an art gallery!

Step inside the space-ship and be transported to a new world, be it down the motorway or up in the hills. If it is physically possible to get a motor vehicle there, then these superb electrically ajustable Connolly hide seats will get you there with a grace and style that other manufacturers have been striving to achieve for the eighteen years since the originals inception.

There have been many changes since the original but these have been minor improvements not wholesale changes.

The lusty V8 engine is still there in an electronically controlled form.

The gearbox (except on the Auto-Only S.E.) has gained a gear and lost alot of noise which was one of the main criticisms with the early cars. The body roll on corners has also been reduced with the revised dual rate suspension.

The engine is more economical and more powerful but a lot of this power seems to be lost to dragging its extra weight, gained with its much needed extra doors, and driving the air-conditioning, larger electrical system and the power steering — which is much appreciated by people parking in tight spaces.

The inclusion of central locking, now on all doors and tailgate saves the tedium of locking all the doors individually and the counter-balanced lower tailgate with clean interior handle and no rust trapping slots, is easily operated by children.

The redesigned grill with horizontal slots hides the radiator and oil cooler and complements the air drawn with revised slots for extra cooling. The revised bonnet hinges and fuel filler flap, help reduce the wind noise as do the electrically operated door mirrors. The original lolypops on sticks may have satisfied the road traffic act but were not much use for rearward vision!

The loss of the front opening quarterlights modernises the styling but now the heating or air conditioning systems are needed to stop the windows misting in damp weather without the noise of a partially open main window.

The sunroof is a beautifully integrated feature which when closed is hardly noticeable from outside or within. And it does not leak!

Structually there is very little difference in the chassis or bodyshell except the shell is now welded together and not bolted as before — which makes light accident damage harder to repair but the shell is still bolted to the chassis on large rubber mounts.

With the original so "right" improvements must be carefully thought out so as not to be just done for the sake of it unlike, it seems, with many of the competition in the executive/leasure market.

But I wonder in years to come if the present new vehicles, with all the electronics and electrically operated systems will still be working with the same degree of reliability as their older bretheren are at present.

Robert Ivins

Continued.

Range Rover badging on the 1970-built model.

No trendy spoiler to catch the kerbs.

Now somewhat battered, but the interior was designed to be hosed out.

Quaintly old fashioned logo.

The power plant doesn't yet fill the engine bay.

The famous protruding hinges.

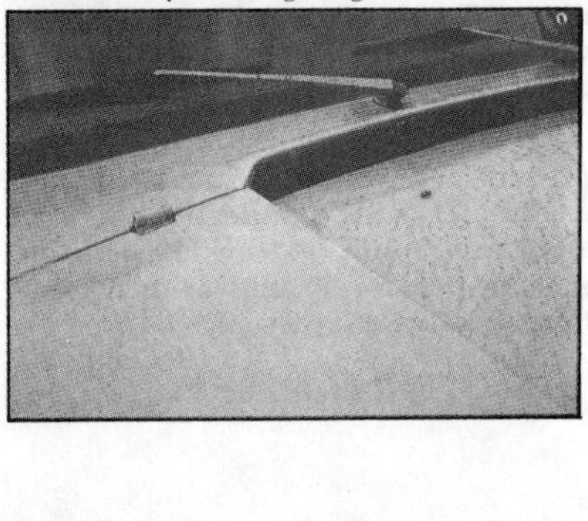

Prototype American-spec rear light.

Then

TODAY'S Range Rover SE, with its new transmission, luxury interior and superb good looks is, for me anyway, the ultimate in motor cars. Its off road performance is superb — the viscious coupling transfer box is a dream — and cruising round town or down the motorway is little short of a magical experience.

If I could afford one, then I would buy one and never want for anything more.

Is today's Range Rover the same vehicle, albeit much improved, as it was first conceived or is it something else entirely? Although today's Range Rover is recognisably the same as 1970 models, it seems to me to be a vehicle miles away from the original concept.

Correct me if I'm wrong, but the Range Rover was first intended as a farmers car. A more comfortable and more refined Land Rover. Something

VITAL S

1970 RANGE ROVER
Engine size: 3528cc
BHP: 132
Compression ratio: 8.13:1
Gearbox: 4-speed with integral
2-speed transfer box
Auto: Not available

Wheelbase: 100″ (2540mm)
Overall length: 176″ (4470mm)
Overall height: 70″ (1780mm)
Overall width: 70″ (1780mm)
EEC Kerbweight: 1724kg 2-door (manua

GVW: 2510kg
Fuel capacity: 18 imp. gallons
Fuel consumption: 18-23 mpg
Tow capacity (on road,
fully braked trailer): 4000kg
Brakes: Servo operated disc all round.
Suspension: Coil springs all round with
Price: £20,000 with a full tank of petro

and now

that could be used around the farm to do all the things the Land Rover could do, but would also suit the family on visits to town.

I don't believe that the company ever thought it would become a cult vehicle. And that, having achieved that status, would become the favoured transport of royals, movie stars, oil-rich Arabs and trendy socialites.

Would anyone who has just spent more than £27,000 on a new SE use it around the farm? Towing a horse box, perhaps. Or a powerboat, maybe. This is the stuff of wealth. Of success. Of status. Of achievement.

Richard Thomas.

TISTICS

RANGE ROVER SE

cc

l

eed with integral morse chain drive
peed transfer box
ed (standard on SE) with integral morse
ain drive 2-speed transfer box
(2540mm)
(4470mm)
1972mm)
1813mm)
kg (4-door) auto.
kg (2-door) manual
kg
imp. gallons
mpg

kg

rear supension.
00

Stick on transfers replace the original aluminium lettering.

New grille, new spoiler

The epitome of luxury motoring.

Tailgate handle now inside.

Choc-a-block with engine and goodies.

Split rear seat backs and carpet everywhere.

What! No hinges!

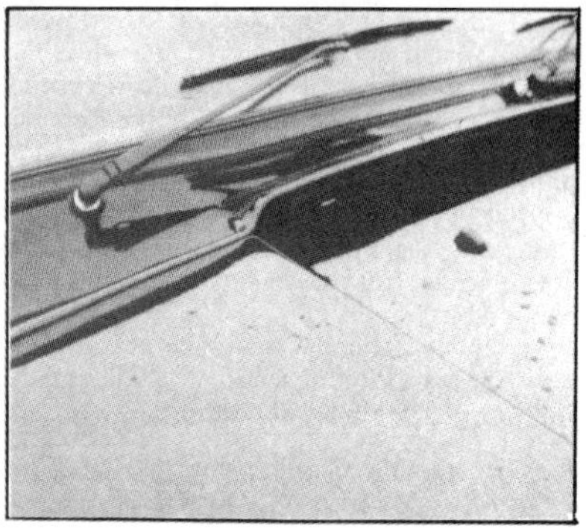

IMPRESSIONS

RANGE ROVER VOGUE SE

● The Range Rover has carved a niche in the luxury car market that it was never really intended for. But the latest changes have made the British 4x4 far more deserving of the luxury image and less of a compromise, as John Henderson discovered with the plush, top of the range Vogue SE.

The old Cornish garage owner and his wife had been very impressed by our bog standard, short wheelbase Land Rover stationwagon, and had wandered around it cooing a strongly accented, 'nice motur that'. The husband told us about the long line of Land Rovers they had owned and then added vehemently, 'we don't like they plushed-up Range Rovers!'

That was about six years ago and our friend would throw up his hands in horror at just how 'plushed-up' they Range Rovers have become. There was some justification in the criticism that the Range Rover was an off-roader with delusions of grandeur: a utility vehicle trying to be a luxury car. It had achieved the prestige almost by accident, but there were still the drawbacks of gear whine and low-speed transmission shunting and clonking. In addition, it had to be handled with due respect for its size.

The transmission noise has been reduced over the years from the original transfer 'box's motorway whine like a cruising Viscount turbo-prop, but it was not completely eliminated by the adoption in 1982 of the LT230 helical 'box.

But with the 1989 model year the Range Rover has been transformed. The heart of the change is a new chain driven transfer 'box with a viscous coupling to lock up the centre differential in slippery conditions. Made by Borg Warner, in Wales, it uses a Morse chain and sprocket drive system. Gearbox noise is reduced to negligible levels: talk to your passenger and you won't hear the faint whirr at all.

Even more noticeable is the absence of low speed shunting and clonking. Our test car was a Vogue SE which has a four-speed automatic 'box as standard. Auto Range Rovers always had less pronounced transmission shunt than manuals, but you could still induce it, particularly in low-ratio, by trickling the car along slowly and lifting off the throttle – as you would in a traffic queue, say. But all that is now a thing of the past.

Selecting low ratio is a lot easier and smoother too, and can be done on the move. The viscous coupling means that you no longer have to engage the diff lock off-road just to be on the safe side. It is all done automatically for you just when you need it. Make a sharp turn from a track to go up a muddy slope and there will be none of the slithering about that used to accompany such manoeuvres with the diff locked, but then the vehicle will lock the diff to climb the slope.

Land Rover say that it takes just half a wheel revolution of spin to get the diff locked, which makes it useful on the road too. It has improved the Range Rover's handling, making it feel much more sure footed. Push it into a corner, keeping the throttle live, and, just as the front starts to come unstuck, the viscous coupling acts and the car drives round the corner instead of understeering: it feels a little like lift-off tuck- in. Respect is still called for however, because if you do lose control you'll soon discover what a big, heavy vehicle the Vogue is!

The Range Rover also feels more lively than before, presumably because of reduced friction in the transmission. It is still powered by the 165bhp Lucas injection version of the 3.5-litre alloy V8, which also develops a beefy 207lb ft of torque. Reckon on a top speed of just over 100mph and 0-60mph in about 12 seconds, with impressive overtaking acceleration for such a big vehicle. But if you give it too much green wellie, you can expect to pay at the pumps: we got 17.3mpg.

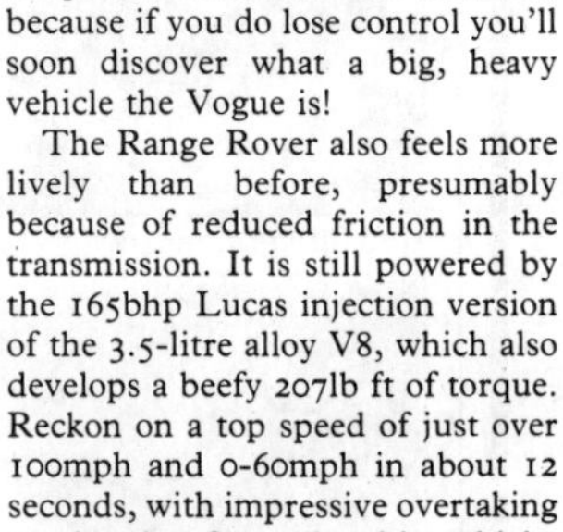

▲

Range Rover Vogue SE still has the 165bhp injection version of the Land Rover's thirsty but desirable all-alloy V8

The 1989 changes also mean the Range Rover has become even more car-like. The electric windows have one-touch opening on the driver's side and there are heated windscreen washers, variable delay intermittent wipers and rear seat latches that no longer require the strength of Goliath to move them. The central locking now includes the tailgate, which also has a button that operates without dislocating your thumb as it did on older Range Rovers.

The heated rear screen is revised to cover a larger area and now also acts as the radio aerial (you always used to forget the aerial was up until you turned into a narrow track and snagged it in an overhanging hedge). Interior changes include new, attractive door trims with polished walnut inserts, the Vogue SE's Clarion radio-cassette player on all models (though the SE has six, instead of four, speakers) and improved heating and ventilation with a quieter, more efficient fan.

The Vogue SE is indeed the most luxurious Range Rover of all. At £28,855 you get air conditioning, the automatic 'box and an electric sunroof that would cost £3613 as extras on the £23,029 Vogue. In addition you get leather trimmed electric seats, colour coded alloy wheels and an SE tag to tell other Range Rover owners that you're one step further up the social scale.

In fact, the Range Rover is now very much a luxury car that is also an off-roader. You no longer revel in the commanding and comfortable driving position and the sound of the V8, while having to make allowances for so many other things. The injection engine produces a fair amount of power while the new transfer 'box at last provides a transmission to match, and the interior and equipment are right up to the standard that the price and the image demanded. ●

Likes: New transfer 'box and equipment levels, handling, character

Dislikes: Fuel consumption

Price: £28,855

The injection engine produces a fair amount of power while the new transfer 'box at last provides a transmission to match

Aristocratic Oiler

Already a legend in Europe, the Range Rover Turbo D holds a few surprises for Aussie off-roaders. Joan Green tried it out.

There's always a sense of privilege when you find yourself in the driver's seat of a brand new Range Rover. The high profile position, lounge room comfort and panoramic view are all part of the unique "Rangie" ambience, as is the muted purr of the big V8 under the bonnet. After enjoying this experience a few times you tend to climb aboard the next test model already programmed with certain expectations.

So it's just a touch disturbing when the noise from the latest model doesn't sound quite right, and still more disconcerting when the vehicle doesn't take off in the usual willing manner even when you know quite well that this is no ordinary Range Rover (if a Rangie can ever be called "ordinary"). The first experience of the 1988 Range Rover "Turbo D" comes as a bit of a shock to just about everyone.

It was quite late in the year when our test model finally arrived at "4x4". Delays, strikes and supply restrictions had brought the long-awaited vehicle into the country without the anticipated fanfare from JRA and it had been difficult to get our hands on one from their obviously limited press fleet. Of course we'd known all about it for ages and had written a couple of stories about it already (after all, it had been on the market in Europe for the past 18 months receiving a great deal of acclaim), so its arrival was no big deal - until we fired it up.

The sound of a diesel from beneath the hood of a Range Rover is one new experience, the sound of this particular diesel is another. This is not the well-known heavy clunking of the old fashioned oiler, it's the new noise of the modern lightweight alloy unit. We'll be adjusting to variations of it from the new generation diesels for some time to come, and a higher pitched, harder and busier hammer is what we must learn to expect.

In Australia the Turbo D is available only in standard form

Having heard that this is Europe's most powerful 4WD vehicle, speedy, fuel efficient and extremely popular, we were expecting an instant take-off. This was the second shock. At low speed the vehicle is decidedly sluggish and won't get you away from the lights without embarassment until you adjust your normal Range Rover driving technique. It takes a little while to get over the astonishment.

Within a day or so however, having learned all over again how to drive a Rangie you can be moving around town much more easily, and positively revelling in the vehicle's performance out on the open road. Criticism wanes as you discover that the Turbo D will perform quite willingly in response to strong encouragement from the right foot.

With diesel fuel being very much cheaper than petrol in Europe it became obvious to Land-Rover some time ago that a diesel version of its popular Range Rover could be a best seller on the Continent. Never one to rush into anything however, the company took considerable care over the selection of a suitable manufacturer for the new engine. The manufacturer finally chosen was the Italian company Stabilimenti Meccanici VM. A further two years were then to elapse before the unit was fully developed and ready for production.

The result was the purpose-built 2.4 litre 4-cylinder turbocharged fuel-injected intercooled engine, VM HR 492 HI, that has set a number of endurance speed records and boosted European sales of the Range Rover to gratifying levels. Whilst its power output is nothing outstanding at 84 kW/4200 rpm, (that explains the culture shock after you've become accustomed to the petrol model's 112 kW at 4750 rpm), torque is 248 Nm at 2400 rpm, and that's the secret of the vehicle's very creditable performance on hilly

highways and off-road. Gearing for the new diesel remains the same as the petrol model.

The Range Rover's engine compartment is normally pretty well packed so it's quite a pleasant surprise to discover this neat and compact unit enjoying more than adequate space beneath the hood. The VM engine features four separate alloy cylinder heads, a triple-K turbo and a cleverly built-in thermostatically controlled engine oil cooler. Everything including air filter and fuel filter are well positioned and easy to reach

Our few days with the Turbo D were sufficient to reverse our initial

Ron Moon drove the "Vogue" version during his JRA trip to England last year

A TURBO IN ENGLAND

We had heard before we left for England on a JRA trip a few months ago that the turbo diesel Range Rover was to be released in Australia. While we vainly tried to get one, my first experience with it was in England on the JRA swan. In fact, it's been my only experience as the turbo Rangie turned up at work while I was away on another trip.

When I first saw that a diesel was being put into this luxury machine and that it was to be a 2.4 litre turbo I could hardly believe my eyes. We're used to four litre plus diesels here and while these machines deliver an adequate supply of power and torque I could hardly believe it could be up to Range Rover standard.

Even when I saw the specs of 84 kw at 4200 revs and 248 Nm at 2400 rpm (compared to the 4.2 litre Nissan diesel of 85 kw at 4000 revs and 264 Nm at 2000 revs) I was hardly shaken by excitement.

Anyway, I finally managed to see one in old Blighty and when I first looked under the bonnet I began to be impressed. It was a lovely installation - tidy with lots of space with intercooler and engine oil cooler to boot.

Firing the machine up the first thing you notice, especially after getting out of a super quiet Vogue SE, is the noise level. It's definitely up a fraction higher, but at idle it's the loudest it gets and it's definitely far from being painful. In fact, it's only a short time before you don't take any notice of it at all.

Driving it came as a pure surprise. Although it didn't lay you back into the seats, the acceleration was brisk once you got the turbo working. It's claimed you can go from 0 to 60 mph or 100 kmh in less than 18 seconds and that's pretty reasonable for a diesel. Certainly its top speed on the open road is a good 90 mph or 145 kph and that's not all bad, especially when you can get better than 30 mph or just 9.5 l/100 km.

On hills you need to peddle the vehicle through the gears a little more than you do with the V8 and that's only to be expected with a motor turning out considerably less than its bigger petrol cousin.

The Turbo D Range Rover was made for the European market where diesel fuel is half the price of petrol. Here in Australia where the price difference isn't so great Turbo D's won't sell as well as their petrol counterparts. However, JRA hopes to take the Turbo D to a level of 10% of its Range Rover sales. There's certainly no mechanical reason why such a figure should not be obtained.

Ron Moon

impression and explain the vehicle's popularity overseas. Far from being a disappointment, this is a very worthy partner for the V8 petrol model and an alternative well worth considering if you're deterred by the petrol Rangie's tendency to guzzle fuel. Even in Australia the extra cost of the oiler (around $4000 more than the standard V8 Range Rover) could possibly be justified by the fact that its fuel consumption is around half that of the petrol version. Although we didn't have the vehicle long enough to prove the theory, the Turbo D's outstanding economy is well enough documented to satisfy all buyers that this is a very frugal beast indeed.

The vehicle's slow get-away was possibly our biggest complaint. The turbo doesn't take effect until above 2000 rpm and gives its best over 2400 rpm, but by keeping up the revs you can achieve a very gratifying and lively performance and quite amazing speed. We didn't attempt to wind up the vehicle to its claimed top speed of 90 mph, but we were convinced that this would have been entirely attainable, conditions and the law permitting.

The only other cause of mild discontent was the noise level in the vehicle we tested - the ear does become accustomed to this after a while, but it was evident that passers-by were aware of the loud exhaust note that contributed to the total decibels inside the cabin.

There are no complaints however about the Turbo D's off-road performance. In fact, Rangie devotees who actually do take their vehicles off-road would be quite impressed by its capability. It can do everything the V8 version can with the same ease and grace, and the added advantage of superior engine braking on the steep downhills.

With the new power plant and its "different" performance of course you still get the superb coil sprung ride, reliable handling, interior comfort and undeniable class of this most elegant of off-road vehicles.

We understand that in UK and Europe you can buy a Turbo D Range Rover with all the fancy trimmings of the model we know as "Hiline" - the only exception being the absence of a manual transmission option. In Australia, so far at least, you can only buy the new diesel in standard form, and this may come as a disappointment to would-be buyers seeking the ultimate diesel-powered status symbol. On the other hand, we can live without the electrical gadgets as well as the Hiline's front "spoiler" (in our opinion the latter spoils the look of the vehicle as well as getting in the way of serious off-road work), and we believe that most "4x4 Australia" readers will agree.

JRA may not be expecting a large demand for the Range Rover Turbo D in this country (we understand that they're budgeting on 10 percent of the total Rangie market), but they could just possibly be in for a surprise.

Standard Interior

Neatly installed 2.4 litre VM engine

Aristocratic Oiler

Life with Range Rover

Through dell, glade, bog, and gridlock. And always with grace.

BY NICHOLAS BISSOON-DATH

• Land Rover's Range Rover enjoys a worldwide reputation for its ruggedness, off-road capability, and myriad creature comforts. An expensive, do-anything go-anywhere machine, the British Range Rover has achieved considerable status among the wealthy. And that's not surprising: who wouldn't be intrigued by the idea of an unstoppable, leather-lined, four-wheel-drive sport-utility vehicle? We certainly were. So to find out if the Range Rover delivers on its considerable promise, we ordered one for a 30,000-mile long-term test.

Range Rovers come equipped with so many standard amenities that the factory offers only two options: leather upholstery and a power sunroof. We ordered both. That brought the price of our Range Rover to $36,450—including the $550 freight charge. We also ordered several dealer-installed options: floor mats, a brush bar, front mud flaps, a dog guard with a shotgun rack (in case we had

PHOTOGRAPHY BY TOM DREW

the urge to repair to a dog show or a hunting blind), a cargo-compartment liner, and sheepskin front-seat covers. These additional features boosted the sticker to $37,859. As we said, the Range Rover is a plaything for the rich.

We picked up our bright-red long-term car in January, 1988, at the port of entry in Baltimore, Maryland. On the trip back to our editorial offices in Ann Arbor, Michigan, the Range Rover impressed us with its comfortable ride, solidity, and highway poise. We also thoroughly enjoyed the commanding view provided by the high seating position.

Our progress was stately, not sprightly. The Range Rover had no trouble maintaining reasonable highway speeds, but the 150-hp 3.5-liter V-8 didn't have enough punch to accelerate its 4397-pound burden with anything approaching dispatch. (That shortcoming has been ameliorated in the 1989 Range Rover—see sidebar.) And despite the vehicle's stability, it displayed an annoying tendency to follow irregularities in the road surface. We had the wheel alignments checked at the 1000-mile service, but the dealer made no adjustments.

February in Michigan means cold and snow, which made our Range Rover one of the most popular rides on the sign-out board. With the aid of its Michelin XM+S 200 tires and full-time four-wheel-drive system, the RR tracked through the worst from Michigan's Mother Nature without a hint of drama. It made a fine long-distance traveling companion, too, providing comfort, lots of luggage space, and peace of mind.

At 3763 miles the brake-circuit warning light came on. The dealer checked the system thoroughly but found no problem. Eventually the light went out by itself, and no further brake-circuit irregularities occurred.

Our Range Rover accrued miles in a hurry. We experienced no major problems in the early stages of our test, but there were a few niggling annoyances. The doors somehow became increasingly difficult to open and close, the ventilation registers and controls malfunctioned, and a rear-seat securing bolt worked itself loose. And just prior to the scheduled 7500-mile service we began to notice an occasional knocking noise from the front end.

Our local dealer (Fred Lavery Range Rover in Birmingham, Michigan—about 45 miles away) performed the 7500-mile service, fixed the minor mechanical nuisances, and traced the knocking sound to loose bolts in the front suspension. That was unsettling, but once tightened the bolts didn't give us any further problems. The service cost us a hefty $214. Strangely, that figure included $62 for repairs that should have been taken care of under warranty.

As winter turned to spring, other problems cropped up. The fuel-injection-system warning light blinked on at the 8708-mile mark, necessitating a trip to the dealer for a new throttle potentiometer. At 11,553 miles, we returned to the dealer with a badly leaking left-front U-joint, loose interior trim pieces, cracked door-handle seals, and a rattling dash. All were repaired under warranty.

With spring in bloom, we took to the hills. The Range Rover continually

New for 1989

Worthwhile improvements to the reigning king of the sport-utility class.

• Our long-term Range Rover earned rave reviews for its off-road talents, roomy interior, and numerous luxury features. We would have liked it more, however, if it hadn't been so slow.

The 1989 edition sees to that shortcoming. The new Range Rover still won't leave a Corvette in the dust, but its bigger, more powerful engine makes it considerably peppier than the old version.

By increasing the bore of the Range Rover's venerable aluminum V-8 from 3.50 to 3.70 inches, Land Rover engineers upped the displacement from 3532cc to 3947cc. That increase boosts power from 150 hp to 178 and raises peak torque from 195 pound-feet to 220.

The additional output noticeably energizes the Range Rover. Maintaining an 80-mph highway cruise is no longer a relentless battle with the wind. And off-the-line performance is dramatically improved. The 1989 model accelerates from 0 to 60 mph in 10.3 seconds, about three seconds quicker than last year's model. Quarter-mile performance improves from 19.4 seconds at 70 mph to 17.8 seconds at 76 mph. And top speed is up a solid 10 mph—to 107. That's enough performance for almost anyone's needs.

Beyond its sheer objective improvements, the 1989 Range Rover is a much more relaxed vehicle than its predecessor. No longer is it necessary to floor the throttle or kick down the transmission to maintain a reasonable pace. Driving is now more dignified. The only drawback is a 1-mpg loss in fuel economy, down to 13 mpg in our tests. For the buyers who can afford this machine, that shouldn't be much of a hardship.

The engine isn't the only aspect of the Range Rover's driveline that's been upgraded for 1989. A new transfer case with a viscous coupling replaces the old manual-lock center differential; it automatically provides progressive lockup as needed. In addition, the new transfer case is quieter than the old unit and is easier to shift from high to low range.

The latest Range Rover also sports a revised interior with updated door panels, a new electrically heated windshield, an upgraded stereo system with six speakers, and more powerful air conditioning. In addition, the '89 model offers a new tan interior color that coordinates with the wood trim much better than the previous model's standard gray (which is still available).

An extraplush model called "the County" has been added to the 1989 lineup. This top-of-the-line Range Rover comes with such standard amenities as a sunroof, leather upholstery, a multidisc CD player, color-keyed wheels, special badging, and richer-looking wood trim.

The County lists for $41,600; the base 1989 Range Rover costs $35,800—a $1400 increase over the 1988 model's year-end sticker. That seems a fair exchange for the many improvements, however. The updates made on the 1989 Range Rover leave it in fine form to battle the new Laforza (see preview this issue) for the top spot in the superluxury sport-utility class. *—Csaba Csere*

amazed us with its creepy-crawly off-road prowess, tackling terrain that at first seemed impassable. And when we were finished playing in the mud, the Range Rover transported us home in comfort.

At the scheduled 15,000-mile service, the dealer fixed a noisy speedometer by lubricating its cable. That problem was fixed under warranty, but the regular service set us back a staggering $353.

The first months of the Michigan summer—usually airy and cool—brought record-high temperatures. With day after day in the 90-degree range, the Range Rover's air-conditioning system got a tough workout. But it performed well. The engine, however, suffered from hot-starting problems on several occasions. At the same time, the Range Rover mysteriously began to pull to the right under braking. And we noticed several new annoying rattles from various trim pieces.

We returned to Fred Lavery to cure the latest maladies. The dealer found that the left-front hub seal was leaking and had spilled grease on the rotor—reducing the left brake's effectiveness and causing the working right brake to pull the vehicle to that side. The fix was a new hub seal and a new set of front brake pads. The dealer also replaced a few trim pieces and rotated the tires. Most of the problems were fixed under warranty; the tire rotation cost us $28.

By the time of the scheduled 22,500-mile service, our Range Rover was suffering from other afflictions. The speedometer was acting up again, and the brakes had begun to screech loudly. The dealer installed a new speedometer cable and a fresh set of rear brake pads. Meanwhile, the service inspection revealed leaks in the power-steering pump and in the rear pinion and differential seals. In addition, the dealer replaced the belts for the power-steering pump and the alternator. Most of the repairs were covered under warranty; we paid $69 for the scheduled service and $158 for the rear-brake job.

As fall approached, our Range Rover remained as popular as ever. But the irritating problems continued. At 24,451 miles, the alternator belt let out a shrill shriek and broke. We replaced it. At 24,736 miles it broke again. We replaced it again. At 24,856 it broke a third time. Because the alternator had shown no such appetite for belts previously, we suspected that the alternator bracket—which had been changed at the last service—hadn't been installed correctly. The dealer confirmed our suspicions and fixed the bracket at no charge.

We racked up the last 5000 miles in

just two and a half months, reaffirming our Range Rover's status as the winter carriage of choice. Despite its popularity, however, our red Rover continued to suffer from niggling maladies. The speedo started to act up yet again, the right-front seat-control switch went on the fritz, and an interior-light switch malfunctioned.

The scheduled 30,000-mile service relieved our checkbook of $570. The dealer also replaced the speedo and the switches under warranty and replaced the fuel-filter bracket—which had been recalled by Range Rover of North America a short time before.

During our 30,000-mile test, we spent $1182 on service and were forced to make five unscheduled dealer visits. Both figures are notably higher than the norm. More positive was our 14-mpg fuel average—a respectable figure considering the Range Rover's weight and our penchant for the throttle.

Although our Range Rover suffered from numerous minor irritations, it never left us stranded in the woods. Off-road or on, the Range Rover transported us in comfort and style. As one logbook entry noted: "This machine is so solid, it feels like a Sherman tank compared with other four-by-fours on the market. It's in a class by itself."

As our test proves, however, buyers who want to enter that class should be prepared to pay a steep admission price—both in time and in money. ●

Vehicle type: front-engine, four-wheel-drive, 5-passenger, 5-door wagon

Price as tested: $37,859 (base price: $33,400)

Engine type: V-8, aluminum block and heads, Lucas electronic port fuel injection		
Displacement		216 cu in, 3532cc
Power (SAE net)		150 bhp @ 4750 rpm
Transmission		4-speed automatic with lockup torque converter
Wheelbase		100.0 in
Length		179.8 in
Curb weight		4397 lb
Performance:	**new**	**30,000 miles**
Zero to 60 mph	13.5 sec	12.9 sec
Standing 1/4 mile	19.4 sec @ 70 mph	19.1 sec @ 72 mph
Braking, 70–0 mph	218 ft	198 ft
Roadholding, 300-ft skidpad	0.65 g	0.67 g
Top speed	97 mph	95 mph
Road horsepower @ 50 mph		24 hp
EPA fuel economy, city driving		14 mpg
C/D observed fuel economy		14 mpg
Unscheduled oil additions		14 qt
Service and repair stops:		
Scheduled		5
Unscheduled		5
Operating costs (for 30,000 miles):		
Service		$1182
Normal wear (tires, brakes, etc.)		$168
Repair		$83
Gasoline (@ $1.10 per gallon)		$2337
Life expectancies (extrapolated from 30,000-mile test):		
Tires		48,000 miles
Front brake pads		29,000 miles
Rear brake pads		24,000 miles

Continued from page 45

ABOVE: Optional leather upholstery compliments wood trim on doors.

BELOW: Ahead of the four-speed automatic trans selector is the Hi-Lo transfer case lever.

Check them out we did. At the Los Angeles area debut, we were given the opportunity to test ten Range Rovers over a 22-mile course in Silverado Canyon, a little south of Orange, in the Cleveland National Forest. The trail used has been closed for some time due to its deteriorated condition. This made it a perfect testing ground to learn what sort of stuff the Range Rover is made of.

The Beast

Off highway is where Range Rover earns its highest marks. Its off-road performance is almost totally belied by its outside appearance. Range Rover will get down and dirty with the best of company and with no complaints. The front air dam is removable to increase the front approach angle and a factory-installed receiver-type trailer hitch does double duty and protects the rear overhang from damage.

Range Rover is full-time FWD with one twist; there is a differential built into the tranfer case between the front and rear propshafts. It works just like a conventional differential, except that it works between the front and rear diffs instead of between two axles. For rough going, this center diff can be locked and the transfer case will then supply power equally between the front and rear differentials. For even tougher going, there is a low range which drops the overall ratio in low range/first gear to 29.13:1. Many die hard out-backers scoff at automatics, but the Range Rover 4-speed works well off road. There is a small loss in downhill braking power with the automatic, but downhill braking in first gear/low range was impressive despite.

The one key to Range Rover's off-highway performance is its suspension. With 8 inches of wheel travel up front and 11 inches in the rear, Range Rover keeps its wheels on the ground regardless of terrain. Keeping tire contact with the ground is a fundamental rule for good off-road performance. Also, a compliant suspension makes the bumps and grinds of the outback a lot easier to bear.

You will hear few complaints about lack of power for off road use. The 3.5 liter V8 makes 195 ft/lbs of torque at 3000 RPM; and since the 2-3000 RPM range is much used off road, the peak torque range is right where it ought to be. The electronically fuel injected engine pulls heartily and happily and feels much more potent at low RPMs than the carbureted models. One added benefit of the EFI is never worrying about flooding or starving for gas when the engine is tilted at a different angle. The engine will run even upside down.

One feature that will enhance the Range Rover's usefulness is its towing ability. Rated to tow as much as 7700 pounds of trailer, Range Rover could handle half the Budweiser Clydesdales with a little space left over for a keg or two of beer. Also, since the chassis is rated to carry over 1600 pounds, there is no need to worry about taking that extra suitcase. The *Hydromat* leveling device keeps the Range Rover level no matter what the load or terrain conditions.

Summing Up

Range Rover's off-road performance limits will probably exceed those of many of its future owners. This is as it should be; Range Rover will then be likely to meet or exceed its owner's expectations. With the rising popularity of the sport/utility vehicle, even in the upper income areas, Range Rover should find a perfect nitch at the top of this class. RRONAs biggest problems will be overcoming the mistrust many Americans have of cars British. Based on Jaguar's recent success at this, Range Rover should be with us for some time to come.□

Range Rover 3·9

The Range Rover has been steadily improved over its not insubstantial lifespan, culminating in the latest 3.9-litre version.

It's hard to imagine a vehicle that produces more polarised responses than the Range Rover. Nobody seems to be blase about the ubiquitous Rangie; you either love it or hate it. The protagonists cheer and point to things like the Range Rover's unparalleled off-road ability, its luxury, performance and status value. The detractors meanwhile seem happy to quote horror tales of unreliability, servicing costs and fragility.

There are, of course, a couple of areas of agreement. Even the most fervent Range Rover hater grudgingly admits the vehicle's superb off-road ability while even the pro-Rangie brigade seems to think the vehicle would be a better one if it was built by the Japanese. All these things are, however, secondary because the Range Rover continues to sell strongly in Australia — a country recognised as something of a sponge for four-by-fours.

If you forget about the pros and cons and concentrate on the purely objective subject of what needed doing to the Range Rover, you'd probably come up with a need for more power. Not that the Rangie was ever a slow vehicle (certainly not by four-wheel-drive standards) but in 1990 and with all the luxury equipment on board a power boost would be welcome.

That's where the engineers concentrated their efforts and the result is a larger version of the existing Range Rover V8. The bore has been stretched to 94mm from 89mm while stroke remains at 71mm, making the engine more oversquare than ever. The new 3947cc capacity gives a 20 percent increase in power with a peak of 134kW at 4750rpm up from the previous engine's 112kW. Torque has also taken a big hike and now measures 304Nm at 3000rpm, up from 257Nm.

Gearboxes have been uprated to cope with the extra grunt and the automatic in our test car, a Vogue (not the bells-and-whistles SE) now has wider faced gears and forged teeth. There is also an inhibitor to prevent over-eager kickdowns from third to fourth at highway speeds.

Brakes are ventilated discs all round and there's now the option of anti-lock brakes for an extra outlay of $3640.

The rest of the package remains pretty much unchanged apart from some worthwhile detail changes to make life less of a struggle. The seats now feature a new foam padding for better support and comfort while increased soundproofing keeps the interior more civilised. New instrument graphics are a big leap forward although the whole dashboard treatment and binnacle is starting to look a little dated.

Ventilation has been improved with better vent placement and the central locking now operates off both front doors. New colors have been added to the range of models along with new interior trim patterns and the option of Connolly hide on the Vogue SE.

If a Range Rover does happen to be your style the 1990 model is going to be more appetising than ever. At $80,130 before on-road costs our Vogue automatic test car is not cheap but undercuts the Vogue SE by almost $15,000. With the 3.9-litre engine, the Range Rover is even more competent off-road, has better on road performance and will cope with towing a bigger cabin cruiser than ever.

CAR

DEBUTANTS

ICELAND SAFARI

Take one Range Rover Turbo D and emergency supplies and you can head for the BIG snowfields, reports Jeremy Crouch, who did just that. But don't forget the Snowcat...

Above: The Range Rover turned in from the coastal ring road along this rough track which led eventually to soft snowfields (right). The landscape (main picture) is bleakly lunar and sheer magic, provided you have no survival worries...

For a moment, time stood still. An eerie silence settled over the glacier. Suddenly, the sky exploded with fire, as the dawn sun, previously simmering behind ice caps, rose brilliant in the sky.

Even Gisli was silent. A few hours earlier, the frank Icelander, without whose snowcat our Range Rover could not have scaled Europe's greatest glacier, had irreverently likened dawn over the Vatnajokull icepack to the appearance of 'a huge fried egg rising in the sky'. Now he sat spellbound like the rest of us, watching his homeland reveal ice and fire in spectacular combat.

Attempting to sum up the scene with words like 'beauty' would have been a waste of time. Like much of Iceland, the Vatnajokull icepack is far beyond that. Colours flow clear and pure through a landscape drawn in bold, distinct lines. And when the power of ice meets the fire of the morning, you feel only a sense of overwhelming respect.

And respect it you must. Iceland is one of the world's youngest slabs of real estate, a place still being sculpted by nature. It is a land perched uneasily on top of the earth's most active and unpredictable volcanic belt, prone to violent eruptions and terrifying earthquakes.

Even under the massive Vatnajokull, dormant volcanoes still simmer, ready perhaps at any time to transform the landscape in seconds. Being there evoked an indescribable feeling; for me, the climax of a journey into the unknown. A journey which had made it possible to experience the closest the earth can offer to driving on the moon – and in fact, Neil Armstrong came here to practise for the 1969 landings. A place where the earth can be hot to the touch, and bubble beneath your feet.

A naturalist's paradise? Most certainly. An off-roader's dream? Well per-

haps. For a start, unless you're an off-roader, it isn't an easy place to visit. The country's interior is only accessible by a four-wheel drive vehicle and then only during a few brief summer months.

Therein, of course, lies its attraction. If you travel by ship from Aberdeen via Shetland and the Faros Islands, 45 long sea hours lie ahead of you. Including stopovers, it means spending four days in the choppy North Atlantic, with at least six more to come on the way back.

One day, when Iceland's beauty has been discovered by the world's holidaymakers as the world's most spectacular destination, it will be different. For now, there's still a spirit of adventure about the journey. Our arrival in the East coast port of Seydisfjour made the national news, for example.

The essential link, the Smyril Line's *Norrona,* sails there only once a week, again only in the summer. And in our case, we were about to subject the Range Rover Turbo D to perhaps its toughest examination ever.

The world's most capable all-round off-road vehicle? Erlingur didn't think so. "These machines are expensive, too heavy and far too complicated," remarked my Icelandic host. "We prefer Toyotas, Hiluxes especially; they are light on the ice, easy to manoeuvre, easy to fix."

I watched his eye turn disparagingly to the standard-fit Michelin X M+S radial tyres. "These won't help you either – you'll need special snow tyres up in the glaciers." But those were unfortunately out of the question, since Range Rover wheels are larger than the common Icelandic sizes preferred by most Japanese manufacturers. It was not an encouraging start.

At least we seemed to have made the right decision in bringing our Turbo D model fitted with its Italian VM engine. In a land where the cost of living is considerable (hamburgers are R21 a throw, while a round of drinks virtually requires a bank loan), diesel is one of the few commodities which remains cheap – a Range Rover fill-up costs around R54. Anyone using a V8-powered petrol-driven version could expect to pay double that.

The inevitable catch is that the Icelandic Government seeks to take that back by imposing a tax of around R108 on diesel drivers to import their vehicles into the country (yes, you must 'import' them). On the other hand, diesel burners can console themselves by filling up from tanks left around the tracks by long departed road workers.

Since population is only possible around the coastal areas, the main ring road which circulates the island is kept in reasonable condition. Apart from the routes around the two main areas of population – Akureyri in the North and Reykjavik, the capital, in the South West – it usually equates to something like a potholed farm track.

Despite the harsh conditions, Iceland is not without its wildlife ——— though it's very different from African varieties. The Range Rover forded fast-flowing rivers of ice water ——— and coped well with the rocky terrain (left). It was the deep, treacherous snow that proved too much for it and made the author thankful for the presence of that accompanying Snowcat (above).

The secondary 'F' routes are really indescribable, their existence indicated by weathered signs on the coastal road verge pointing blankly toward a distant icy wilderness. To European ears, the inscriptions are meaningless – 'Eyjafjallajokull', 'Jokulheimer', but to Icelanders, 'Island of the Ice Glacier' and 'Entrance to the Glacier' tell their own story. Standing on the windswept Torfufell soon after our arrival – higher in the Icelandic wastelands than a jet circling over Heathrow – I began to wish I'd read the book.

The Range Rover was stuck fast in a deep snowfield and the arrival of casual assistance seemed about as likely as finding a friendly Land-Rover dealership around the next peak. The original objective had been to cross the Sprengisendur, the infamous sandy track which links the north and south Icelandic coasts, in a single day, a distance of well over 1 600 km.

Our initial progress had been encouraging, despite the steep slopes of the rocky routes which seemed to have been constructed to find the breaking point of the Range Rover's coil sprung suspension.

My mental picture of an attractive receptionist awaiting our arrival at the Hotel Edda several hundred kilometres south was becoming clearer by the minute. It vanished abruptly as we bumped over yet another volcanically formed crest to find the track covered in a thick white blanket.

Nothing for it but to lock the differential and push forward. The snowfield didn't look deep and I felt sure that the Range Rover would be able to cope. I was wrong, and what was worse, I knew it long before we ploughed to a halt, both axles firmly embedded in snow; yet to stop would have been fatal.

It took us three hours to raise the axles clear of the snow by jacking-up each wheel in turn and placing a bed of stones beneath the tyres. Weeks later, back in England, it took almost as long to edit our Mitsubishi video footage to remove some of our more colourful comments...

We were beginning to appreciate the necessity for the kind of thorough preparation so strongly recommended by the Icelandic authorities. Our plans had been laid under the guidance of polar travel experts, Arctic Experience who, from their British headquarters in Banstead, Surrey, advise increasing numbers of tourists headed for arctic regions.

Traversing the interior during July or August should be no problem, we were told; we'd need a full spares kit, plenty of jerricans for fuel, camping equipment with survival rations, thermal clothes and, of course, a first aid kit. To get the most out of the trip, we'd also packed a video camera.

Having assembled that little lot together with a spade and a high-lift jack, we'd been surprised to find ourselves sparsely equipped, in comparison with our fellow ferry travellers whose kit included everything from sand ladders to two-way radios. They had seen the Vatnajokull...

Tackling the largest glacier between the Arctic and Antarctic circles with a standard Range Rover on tyres more suited to avoiding traffic patrols on the M1 sounds foolhardy. Indeed, it *was* foolhardy, or at least, it would have been, if we had not enlisted the help of a Snowcat to accompany us. With such powerful assistance, there could, I felt, be no problem. It wasn't long before I began to have my doubts.

Vatnajokull has tongues of creeping ice stretching down almost to the south coast and it was one of these, Skalafellsjokull, that we had chosen to gain entrance to the icepack. But since that ice pack sits around 1 500m above sea level, it was soon clear that merely reaching the glacier would be a feat in itself.

Certainly it had seemed that way when, with snowcat pilot Gisli Hjalmarsson aboard, we had started up the steep track from the coast. "This is the worst road in Iceland," he had observed unnecessarily as I shifted the Turbo D into low range. "The bulldozer hasn't quite finished here yet."

To Rover's considerable credit we managed eight kilometres of the mountain track before sliding to a halt at the edge of the snowfields. "Shove it in low range third and see how far you can get," Gisli suggested helpfully, jumping safely clear.

The result was inevitable, as it would have been for any conventional four-wheel drive vehicle. Seconds later, the Range Rover was stuck deep in Vatnajokull's clutches. No matter. Avoiding the ridiculously positioned plastic stoneguard on the Range Rover front end, we attached a tow rope. Our 'go anywhere' machine was soon slithering behind 90 kW of Snowcat.

Once we reached flatter ground, Gisli unhitched us and took over the steering wheel to demonstrate his snow driving technique. The secret, it seemed, was to select low range in the transfer gearbox (*contrary* to the vehicle handbook's recommendations) and to move the wheels forwards as slowly as they would turn, letting them roll back the instant they encountered any obstacle against which they might spin.

Despite knee-deep snow, he was gradually able to create a path with the front wheels which the rear wheels could run into. Once I mastered the technique I was able to negotiate all but the steepest slope without the help of the snowcat.

Lesser, lighter vehicles might have matched the Range Rover there. None could have equalled its all-round terrain capability, displayed on the rocky climb up. Even more impressive was its wading ability. The interior tracks are intersected at frequent intervals by deep, fast flowing rivers whose beds are often sandy and deceptive. Initially we approached every one with some caution, pulling on waders and splashing elaborately out into the middle with depth testing poles.

It wasn't long before we realised that the Range Rover would manage almost everything up to headlamp height with consummate ease. Change down to low range 2nd or 3rd, select a suitable crossing place avoiding the tracks of other vehicles (which may have crossed earlier in the day when the depth was lower) or the narrowest point (where the current will be at its fastest) and straight over.

Simple. It's easy to get over-confident however, and vehicles are frequently swept down river with fatal results. You soon learn not to take on the elements but to respect them.

If you are hit by one of the frequent sandstorms, the only option might be to retrace your steps to one of the remote but welcoming warden huts scattered throughout the interior. It might last an hour, it might last a week. Similarly, if the river you want to cross is too deep, you simply wait until the level is lower.

It's natural obstacles like these which make estimating journey times across the interior impossible. A given trip might take a few hours, or a few days, or it might be impossible. At one point, we were stopped by a group of Icelandic officials who presented us with a ticket – and proceeded to photograph us and the Range Rover.

They explained we were about to enter a zone which seismographic readings had indicated was about to undergo a major earthquake. There would be another official on the other side to collect our ticket. If we didn't reappear, they would at least know who to look for. Though it seemed almost suicidal, some locals were driving past as if this was an everyday occurrence. Iceland provides an unforgettable driving experience for its roads are a convenience of secondary importance which is never allowed to disfigure the landscape. The Icelanders feel no impulsion to improve them, fearing perhaps that if all their country's beauty areas were easily accessible, their attraction would be sullied. Perhaps there's something to be said for that.

We once left the Range Rover to climb to the top of the active volcano Kverkfjoll, reaching the summit six hours later. There, we felt the ground bubbling beneath our feet as we walked among steaming lava beds situated just metres away from a frozen ice lake.

There's also the option of travelling even further north. I tried one of Icelandair's superb one-day tours to the Greenland island of Kulusuk above the Arctic Circle, where the snow is over five metres deep and the Eskimos live in poverty of Ethiopian proportions.

It's not cheap, but then, neither is the cost of visiting Iceland in the first place. From Aberdeen in Scotland, you catch P&D's comfortable *St Clair* to the Shetland capital Lerwick in order to link up with Smyril Line's stylish *Norrona* for the journey North via the Faroe Islands. From Europe, you catch the *Norrona* direct from the Danish port of Hanstholm. Either way, the return cost for four people and a car will equate to R5 400; better, perhaps, to fly and rent a 4wd vehicle there.

After 16 days and 4 000 km, the Range Rover emerged with its manufacturer's claims intact. We were able to keep up higher average speeds than its Japanese counterparts could manage in driving conditions that were usually atrocious, often abysmal. With a little less weight and the right tyres, it would have been almost unstoppable.

And Iceland itself? What better than to quote the words of one of the first Europeans to visit the country:

"Here, everything is pure and untouched as the hand of the creator left it. Mother Earth needed but few expedients to accomplish this wondrous piece of workmanship. Only sky, water, cliffs and her own inner fire."

Two hundred years later, the fire still burns... ●

Where the latest viscous coupling 4WD system for the 1989 Range Rover with manual transmission is concerned, it's good news and bad news.

First the good news.

Most all-wheel drive vehicles are likely to go over to viscous drive couplings in their transmissions eventually. There's a great deal to be said for their reduced mechanical loss, reduced operating noise and their light weight. Until now the complications involved in transfer gearboxes and centre differentials have been accepted because 4WD vehicles are a relatively small proportion of the world car park. The advent of increasing numbers of 4WD passenger cars using viscous couplings however, has made them more cost effective overall, and thus a viable proposition in the off-road market as well.

The coupling has been adopted for passenger cars because of its simplicity of operation which requires no action on the part of the driver. While it's all very well to expect an off-road expert to understand the nuances of drive selection, central differential locking and so on, manufacturers have always been wary of letting "Mr Average" loose with such complication, despite the very real safety advantages.

Viscous couplings comprise numerous plates in a row, immersed in a thixotropic silicone substance. The idea is that, as with a viscous drive cooling fan, the temperature-sensitive substance takes up the drive when it reaches a certain temperature. With input and output shafts turning the plates at approximately the same speed, the temperature is kept relatively low. The moment the input and output speeds vary however, the substance in which the plates are immersed heats up and begins to lock the shafts together. Because the friction medium is a silicone derivative in shear between the plates, there's no wear involved.

More importantly, drive is taken up automatically. It can also be directed to the front or rear in any proportion, depending on the design of the coupling. Furthermore, because the viscous action takes up the drive automatically, the vehicle can be configured for two wheel drive the majority of the time, with four wheel drive coming into play only when the normally non-driven wheels begin to spin. Generally though, it is considered more beneficial for the system to remain in constant four wheel drive with a variable front to rear power bias, depending on the circumstances.

Though viscous drive couplings are presently seen more in passenger cars, they are becoming available as aftermarket conversions which ultimately are likely to make limited slip and locking differentials obsolete. The technology which these components represent is advanced. On that basis alone, it seems odd that "Aunty" Rover should have beaten the Japanese to the punch!

In the case of the Range Rover, the coupling is used in conjunction with the latest Borg Warner chain driven transfer box. This has transformed the vehicle into something that's as foolproof as any 4WD can be. The idea is to improve still more the Range Rover's appeal to drivers with limited off-road experience — for which you can read the "Toorak Taxi" set!

Forgetting the driveline for the moment however, the bad news is that all the criticisms of this vehicle we've published in the past remain more or less pertinent.

Let's just recapitulate on a few of them.

For off-roading there are far too many hung-on bits which are easily damaged under even relatively mild conditions. We tend to knock the corners of the rear bumpers off every time we leave bitumen!

Those Michelin tyres, excellent as they are on bitumen and gravel roads, are woeful as soon as the terrain begins to challenge.

The cabin of the 1989 model has been updated as a matter of course, and there are a number of improvements, but there are still plenty of the old annoyances. It may sound like carping, but the points in question are important from the ergonomic aspect.

First there's the ignition switch. Not only is this mounted at 90 degrees to the driver's sight on the control column shroud, but it is also

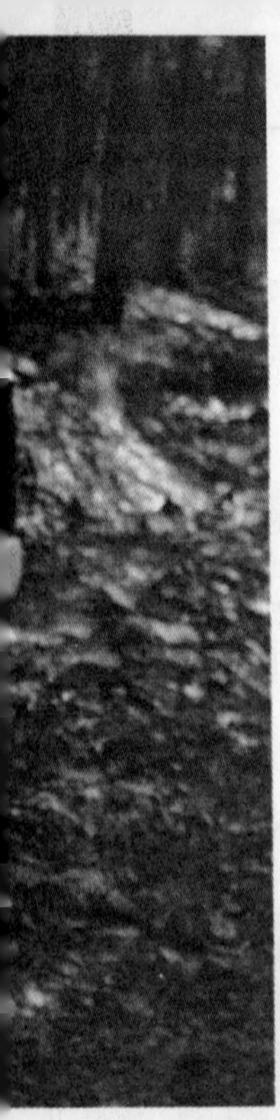

FLUID DRIVE TEST DRIVE!

We recently took a couple of the latest viscous drive Range Rovers for a work out on some of our toughest hills. Here's our thoughts!

located in an indentation. Almost impossible to see when seated at the wheel, we found it necessary to insert the key before climbing aboard in order to avoid frustrated fumbling.

Then there are the control column stalks. Because Range Rover puts more emphasis on sales in continental Europe and the USA, these are set up for left hand drive conditions — the indicator on the left, with the windscreen wiper controls on the right. OK, we can get used to that after a couple of days, but the windscreen wiper activation itself is also an annoyance. Wiper speeds are controlled by moving the stalk upwards instead of down. At least the actuation comes good once you swap the stalks over from right to left!

One improvement however, is that the wipers now have an adjustable variable delay feature, together with heated windscreen washer nozzles for operation in freezing conditions.

The electric window controls mount in the centre console armrest, but facing forward on an angle. To see which switch you're operating, you almost have to crane your neck forward and then twist! Not only that, but the lower, forward mounted switches operate the rear windows, while the rearmost, higher placed switches operate the front windows!

As for the rest of the controls, well, we challenge anybody to try and make sense of the heater, demister, and air conditioning controls. We never did! In fact the whole environment control centre is new and improved in terms of its actual function. At least that's what they tell us. We can't confirm it!

On the bright side, although the seats are mounted so high that it's often necessary to bend the head to look through the windscreen, they are supremely comfortable. Mind you, it takes a while to get them positioned just so. There's a seat cushion angle adjustment, together with a wheel controlled backrest adjuster which really has to be operated from outside the vehicle with the seat unoccupied. Otherwise you can wind forever and nothing seems to happen.

There are new inner door panels with polished wood inserts together with flush mounted door locks. The plastic panels have molded armrests, together with carpet trim and larger door pockets.

There are push button rear seat latches, higher quality audio speakers and a heated windscreen. There's even an automatic headlight control which switches them off when the ignition is switched off.

But how capable is this latest evolution of the Range Rover? Has its off-road capability been improved?

In a nutshell, yes.

Off-road the viscous control unit comes into play automatically, maintaining a minimum torque and speed variation between the axles. This makes it a very smooth handling vehicle with a continuous thinking

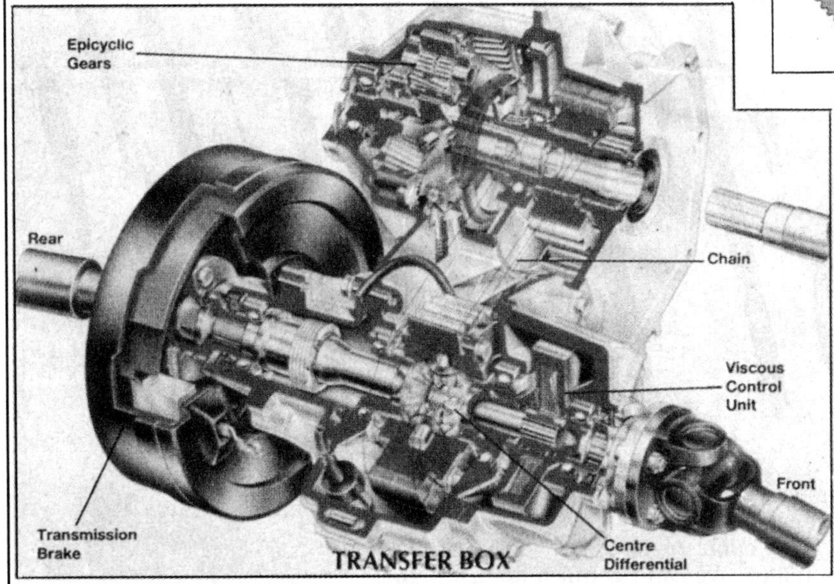

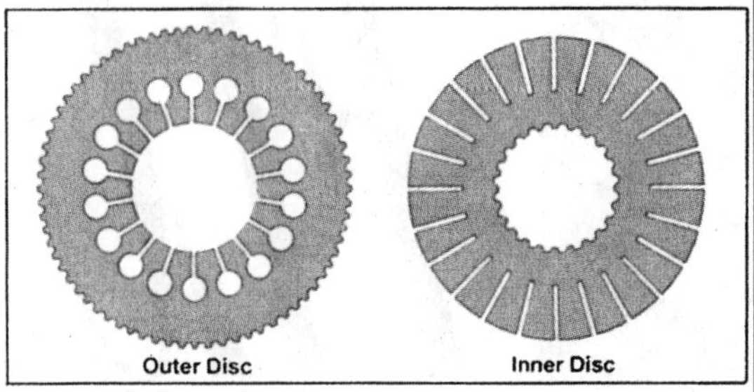

Transfer box showing the Viscous control unit left; right is the viscous unit showing the drive discs and top the outer and inner discs.

function which allows it to adapt to all conditions. This means the driver can concentrate on the terrain, leaving the rest to the transmission system.

Only under very wet conditions was there any sensing of the viscous lock up when driving on the bitumen. It could only be noticed when it was purposely induced, remaining completely unobtrusive at all other times.

We were particularly interested in trying the Rangie on our infamous test hill. It's one of those challenges which brings butterflies to the stomach whenever confronted by it for the first time after a number of weeks away. Steep — around 40 degrees in some sections — it's a mass of rocks and boulders, with a gooey clay underneath. In places it's deeply eroded with 12 inch deep gullies etched out by the streams of water which pour down when it rains. Making it worse on this occasion were some foot high rock steps which had appeared since the last time we'd used the hill. Those, and a couple of days of heavy rain earlier in the week, made the butterflies feel more like wedgedtailed eagles!

It's best not to pause too long at the bottom. The imagination can tend to get things out of perspective. Into low second then and at it.

The Rangie ran like a champion. It bucked and bounced its way up, clawing at the wet clay and rocks. You could feel the imminent onset of wheelspin, only for it to evaporate almost before it was felt.

There was a momentary hesitation when it reached the worst steps with the deepest erosion, but a blip on the throttle bounced it up and over with only a minor admonishing 'clunk' as something underneath met mother earth. Then it was all plain sailing for the remainder of the kilometre-long climb.

The descent was a breeze. Again in low second, the Rangie ticked its way down, dropping hard over the step, but never with any suggestion of either end of the vehicle getting away. On reaching the bottom it was good to breathe again!

Once the hill has been conquered the first time, its perils seem tame. So, after a turn around, we were off on the climb a second time.

Now however, the initial run had chewed the clay to a porridge in places. The rocks were liberally coated with the stuff and it was soon obvious the vehicle was making far heavier weather of it. Where previously there had been strong grip, now there was some slipping and sliding, promising that the step would be interesting to say the least!

It was.

In fact, no matter what we tried, we couldn't get over it. Perhaps if we'd reversed all the way to the bottom of the hill and tried again, we might have been able to maintain enough momentum. As it was, each time we dropped back a few metres and gave it another go, wheelspin at all four wheels prevented any worthwhile gain in velocity sufficient to jump the rocks.

Undoubtedly with more appropriate tyres the Range Rover would have done better. Under those conditions however, we decided to give the hill best as it would have been only too easy to damage the unit. Damage to Rangies is expensive to fix.

It's no real disgrace for a vehicle to fail in the climb, even in the dry. In the wet it's entirely forgivable. In this test the hill was by far the hardest feature to which the vehicle was subjected. The rest of the time it coped with the usual rough terrain

Power was transmitted smoothly even on the steepest slopes.

You'll be seeing viscous control units in more and more makes of 4WD's.

FLUID DRIVE TEST DRIVE!

and eroded gravel roads without a sign of trouble. Indeed, under these wet conditions it was wholly confidence-inspiring, both in terms of road holding and handling generally.

There's no doubt in our minds that viscous couplings are the way to go in the future. Not only do they obviate one more complication associated with difficult off-roading, but they also operate more efficiently than traditional locking differentials.

One wonders how long it will be before Toyota and Nissan claim to have discovered this new magic!

TURBO

The Turbo D is just as quick along twisting bitumen as a V8 Rangie, and while performance is similar, fuel consumption is not.

THE NEEDLE of the red Range Rover Turbo D's speedometer was indicating bang on 160 km/h when the white NSW Police Falcon flashed by on the other side of the divided carriageway. The posted speed limit was a mere 100 km/h. At that speed in fifth gear the Turbo D's little exhaust-driven compressor was only just beginning to do its thing.

Honestly officer, the temptation to answer the question that is as old as the motor vehicle itself 'What'll it do?' – proved too much.

Perhaps the Policeman Plod in the Highway Patrol car didn't have the gutter-mounted radar unit switched on, for the Falcon didn't throw a U-turn across the median strip. At least not tha I observed. Then again, the radar might have flashed up some highly-illegal numbers and the officer failed to find the two-tonne, fire-engine-red offender as it skulked guiltily in the backstreets o the next town.

It took a short downhill stretch to wind the Rangie up to the old imperial ton. On the flat it simply wouldn't break through the 150 km/h barrier. Slow the Turbo D isn't the turbocharged and intercooled 2.4 litre diesel engine, made by the Italian firm c VM, is a pint-sized Hercules.

Its been matched with some pretty smart gearing too, which allowed full advantage to be taken of the motor's 84 kW and 248 Nm power and torque peaks. Those figures are produced at 4200 rpm and 2400 rpm respectively.

The effect of the KKK turbo was

FIRST TEST

Noisy, smelly and slug-slow? Nope. Sensible is probably the word which best summarises the viscous-coupled Range Rover Turbo D.

It's even fun.

Eulogy by JOHN CAREY
Images by LEON FAIVRE

D-LIGHT

Although the Turbo D is only available in standard trim, there is no way it can avoid a 'luxury' tag. Turbocharged and intercooled 2.4-litre diesel engine (below) returned 12.2 litres/100 kms (23.2 mpg).

The effect of the KKK turbo was noticeable from around 2400-2500 rpm, depending on engine load. That meant peak torque and the beginning of the turbo's tune occurred at pretty much the same time, which was at 100 km/h in fifth. Turbo effect in fourth kicked in usefully at 80-85 km/h, perfect for open road overtaking.

In fact, it was possible to drive the Turbo D so that the only time the compressor wasn't pumping was during initial acceleration from a standing start. Changing up a little beyond 4000 rpm always dropped the tachometer needle into the meatiest part of the boost zone.

From experience I know a Turbo D is just as quick along twisting bitumen roads as a V8 Rangie. While, in practical terms, the performance is similar, fuel consumption is not.

The Turbo D consumed an average of 12.2 litres of distillate for every 100 km travelled (23.2 mpg for diehard imperialists). That's significantly better than the 20-25 litres that petrol-propelled Rangies commonly return, and frugal enough to endow the Turbo D with a range between fills of 650 km or so.

The Turbo D has its drawbacks too. The diesel rattle doesn't have the same aural appeal of the ear-tickling V8. Despite the obviously thorough job Land Rover has done with the application of sound-deadening material the Turbo D was always more noisy than any of its V8 brethren. The vibes at idle produced by the four-cylinder diesel were also noticeable inside the sumptuous cabin.

Although the Turbo D is available only in standard trim and with the five-speed manual cogbox, there was no way it could avoide a 'luxury' tag. The multi-adjustable cloth-covered seats, first rate rad-cass unit, electric glass front and rear, plush carpet and so forth saw to that.

The fast glass is a new feature in standard Rangies, introduced at the same time as the chain-drive transfer case and viscous coupling. Those like myself, with a natural aversion to unnecessary electric motors and circuits, won't applaud the decision to endow every Range Rover now available on the Australian market with the damn things. Cheaper Range Rovers are the ones most likely to actually be used off road, where the electric windows may be affected by dust and water.

The new transmission was as impressive in the Turbo D as I'd found it in the Vogue auto tested for the May issue of *Overlander*. Since the Turbo D was a manual, the easier gear swapping offered by the almost backlash-free chain-drive system was appreciated. Some modifications to the gear selection system seemed to have done the job too. The shift quality of the box was first class.

Most other things about the Range Rover are also first class. The Turbo D has an even lower first-gear ratio than its petrol-powered relatives. Its total low-range reduction is a shade over 42:1, one of the most impressive figures around. That number helps the Turbo D scale impossible looking inclines with arrogant ease and descend cliff faces at walking pace.

The long-travel suspension that's been the Rangie's hallmark ever since it was launched, nigh-on a score of years back, keeps the wheels on the deck and driving when off the road, and gives a bum-cossetting ride on dirt roads and blacktop in the Turbo D.

And, as has been the case for the past few years, the build quality and fit standards showed yet another step forward with the most recent update.

All of which makes the Turbo D a most sensible means of conveyance. Fuel efficiency, range and performance, with a slick and smart transmission, spiffy cabin and good finish make it the right choice for the Range Rover purchaser with serious work in mind. The only question which remains to be answered is simple. Is there any such thing as a Range Rover buyer who honestly intends to use the vehicle's 4WD ability and who will appreciate this sensible Turbo D?

brake dancing

John Beese examines the latest spec Range Rovers and finds them extremely stoppable

New engine specifications
3.9-litre V8 petrol

Bore	93.98mm
Stroke	71.12mm
Capacity	3947cc
Valve gear	Hydraulic tappets ohv
Comp ratio	9.351
Fuel injection	Lucas L – Jetronic p
Max power	185.1bhp @ 4750rpm
Max torque	235.1lbf.ft @ 2600rpm

2.5-litre turbocharged intercooled IDI diesel

Cylinder layout	4 cylinders in-line
Bore	92.0mm
Stroke	94.0mm
Capacity	2500cc
Valve gear	ohv
Comp ratio	22.51
Fuel injection	Bosch VE indirect
Max power	119.4bhp @ 4200rpm
Max torque	209lbf.ft @ 1950rpm

1990 Range Rover prices (UK retail)

Turbo D	£23,784
Vogue Turbo D	£25,905
Vogue (manual)	£25,506
Vogue (automatic)	£26,830
Vogue SE	£31,949
ABS option	£1,249
Catalyst option	£450

This manoeuvre would have had Roger Crathorne's eyeballs out on coil springs. Land Rover's demo driver supremo would probably have blamed the rear-light electrics for playing nasty tricks had this not been the latest spec Range Rover Vogue SE fitted with anti-lock brakes (ABS) developed by Land Rover in conjunction with WABCO. Descending a steep, slippery and unevenly banked slope into a gulley we dipped the clutch and jammed the brakes on. Throwing the off-road rule book out of the Vogue's electrically-operated driver's door window didn't come naturally after years of selecting the right gear and track for a descent and then allowing the vehicle to lower itself down gently under engine braking.

The fact that the vehicle we were testing is priced at just under £32,000 was also playing havoc with the senses. But in the interests of *OR&4WD* magazine it had to be tried. To be fair the bank we had chosen had ample run-off space and no damaging rocks or trees. In the event we could almost have completed a Kitzbühl ski-slalom course, so good was the steering control under braking. We also tried braking at speed on a narrow, curvy gravel track and, again, the steering control under full braking was remarkable. We were braking hard in situations where normally you wouldn't put your foot near the brake pedal for fear of losing control, with the only real options for retarding speed being to back off the accelerator and maybe drop a gear.

The system, which is fully electronic and operated hydraulically, is described as a four-channel system, which means it applies the ABS braking principle to each wheel independently. It is this fact that allows the system to function on a permanent four-wheel-drive vehicle and work in all situations. It compares favourably with the system now optional on the G-Wagen, which doesn't work off-road, and such systems as that fitted to Vauxhall's Cavalier, which rely on disengaging the drive line to the rear wheels, allowing the front and rear axles to work independently. According to Land Rover, under braking the system monitors each wheel's state of grip 250 times per second, a claimed two-and-a-half times faster than any other 4×4 competitor. Using this information the degree of slip at each wheel is then adjusted up to 10 times per second. When the brake pedal is applied the system operates conventionally until it detects that one or more wheels are slowing too quickly and approaching a lock-up situation. It then closes the valve in the hydraulic supply line to the wheel and releases the braking pressure to that wheel by opening a second valve and 'dumping' the hydraulic fluid back to the reservoir. It would be completely wrong to say that this system changes the technique you should use when driving on or off-road – it won't even stop the vehicle in a shorter distance. What it does do, however, is provide a fool-proof back-up to panic braking in a hazardous situation, giving the driver the maximum possible steering control.

All petrol model Range Rovers now feature the 3.9-litre fuel-injected V8 engine, a re-worked version of the trusty 3.5 unit. The unit, which develops 185.1bhp (a 12 per cent increase over the old unit), was

● **The ABS was particularly effective wher**

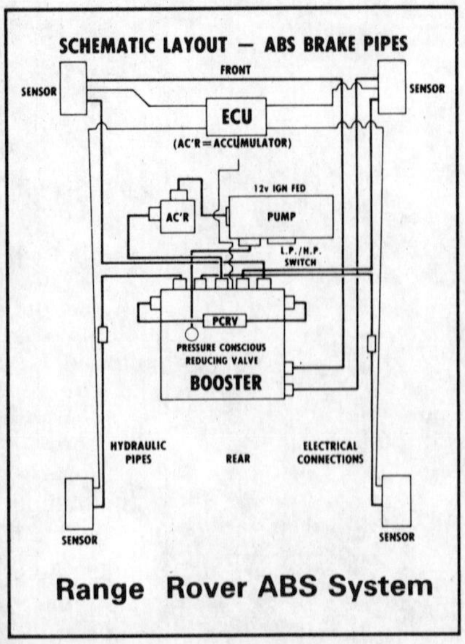

Range Rover ABS System

loose gravel surfaces

designed for export models of the Range Rover going into countries that demand power-sapping catalytic convertors. The UK spec models have the additional power benefit with the convertor only available as a £450 option, which won't please the greens. Certainly the V8 was extremely spritely on our extensive road test, but as the 3.5-litre unit wasn't short of power for normal on and off-road duties, presumably the extra power will appeal to those requiring the vehicle to tow heavy trailers. The automatic gearbox has been strengthened and modified to prevent unnecessary kick-down to third gear during motorway cruising. The Range Rover Turbo D engine increases in size from 2.4 litres to 2.5 litres, with a claimed increase in torque from 183lbf.ft to 209lbf.ft. The diesel was noticeably quicker and able to keep pace more effectively with stop-start fast-moving motorway traffic. We hope to be testing this unit off-road and towing a heavy trailer in the near future.

TEST EXTRA

Range Rover Vogue SE

Anti-lock brakes and a 3.9-litre V8 push Range Rover's Vogue SE above £30,000 but it's worth every penny

Price *£31,949* **Top Speed** *108mph* **0-60** *11.3secs* **MPG** *14.6*
For *Supreme on and off-road ability, sumptuously equipped*
Against *Even thirstier*

Top speed of 108mph and luxury saloon comfort make Vogue SE unique. Predictable chassis with fine grip but pronounced roll and strong understeer when cornering hard. All-alloy 3.9-litre V8 develops 185bhp with fine torque spread

THE RANGE ROVER'S CURRENT glossy advertising campaign says it all. Above a picture of a mud-caked throttle and brake pedal is the line: "After 20 years, two radical improvements to the Range Rover." The small print explains: for 1990 the Range Rover Vogue gets the more powerful 3.9-litre V8 and the option of advanced four-channel anti-lock brakes.

Topping the price list at a hefty £31,949 is the Vogue SE, up £2228 from the old 3.5-litre SE. But included in the price are the anti-lock brakes and the *Autocar & Motor* award-winning Triplex heated windscreen, which is optional on the rest of the range. As well as developing significantly more torque and an extra 20bhp, the engine now runs on 95 octane unleaded as standard and there's a £450 three-way catalytic converter option. On sale in the US in catalysed form for over a year now, the 3947cc all-aluminium alloy V8 gets its extra capacity from an increase in bore from 88.9 to 94mm. With a new free-flow exhaust system rather than the three-way catalyst, the UK-spec 3.9 develops 7bhp more than the US version for a peak of 185bhp at 4750rpm. Maximum torque is up 14 per cent over the 3.5 to 235lb ft at a usefully-low 2600rpm, with 90 per cent of that figure available all the way from 1500rpm to 4500rpm.

In SE form the Range Rover drives through a ZF four-speed automatic gearbox. Internal ratios are unchanged, but there is now a speed-related inhibitor that stops it kicking down to third above a certain speed — full throttle on the old model would always cause the 'box to shift down even at top speed. And, for the first time, SE buyers will be offered the choice of the five-speed manual box, now with wider gears, forged dog teeth and an oil cooler with its own pump. It's a no-cost option on the SE though, with no chance of a refund of the £1324 extra charged for an automatic gearbox in the regular Vogue.

With permanent four-wheel drive, a very 'stiff' viscous coupling across the centre differential and the requirement for it to be equally effective off-road, the four-channel anti-lock had to be a very advanced system indeed. Developed over five years in conjunction with Wabco, supplier of the anti-lock for the Porsche 959 and Carrera 4, it uses an electrically-powered hydraulic booster pump to provide the servo effect.

A supply of brake fluid at high pressure is distributed to the calipers via four independent brake circuits by an electronically-controlled valve block. By comparing the input from conventional wheel-speed sensors at each corner, the control system senses impending lock-up at any wheel and bleeds the brake pressure from that paricular circuit back into the reservoir. The key to its successful operation on a Range Rover is both this four-channel layout (most anti-lock systems have only a single brake circuit for both ▶

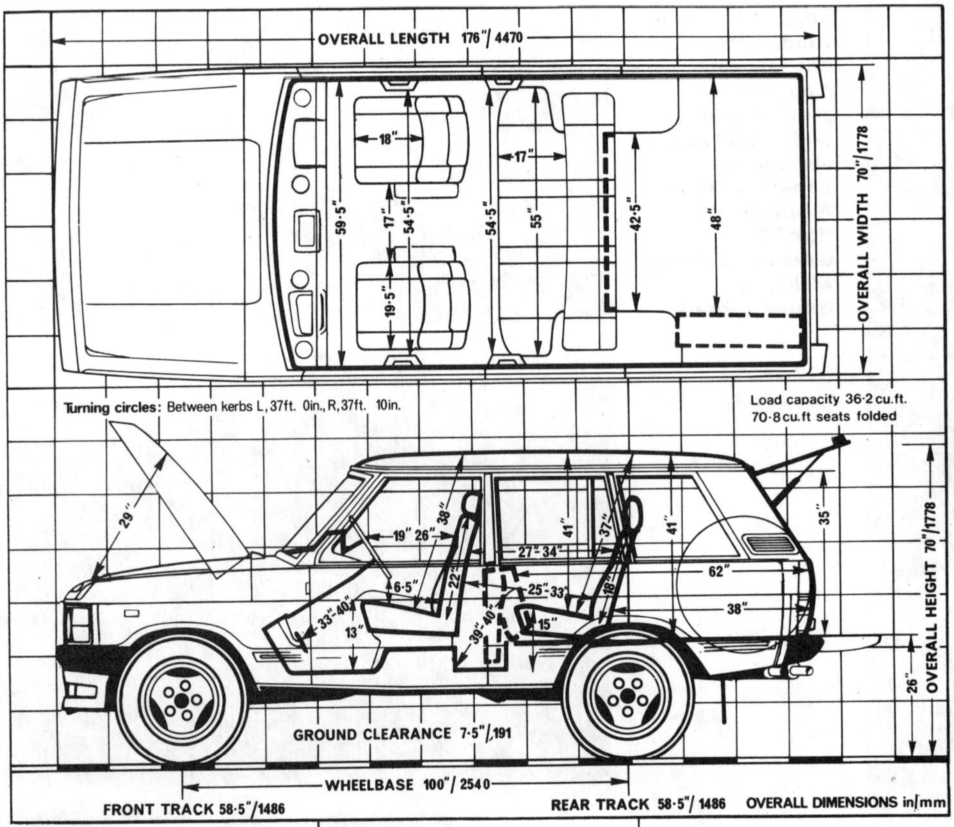

PERFORMANCE

MAXIMUM SPEEDS

Gear	mph	km/h	rpm
Top (Mean)	108	174	4000
(Best)	110	177	4100
3rd	108	174	5500
2nd	73	117	5500
1st	43	70	5500

ACCELERATION FROM REST

True mph	Time (secs)	Speedo mph
30	3.9	31
40	5.8	42
50	8.2	53
60	11.3	64
70	15.4	74
80	21.3	85
90	31.5	95

Standing ¼-mile: 18.5secs, 75mph
Standing km: 34.7secs, 92mph
30-70 Thro' gears 11.5 secs

ACCELERATION IN EACH GEAR

mph	Top	3rd	2nd	1st
10-30	—	—	—	2.9
20-40	—	—	—	3.5
30-50	—	—	5.2	—
40-60	—	—	5.7	—
50-70	—	—	7.3	—
60-80	—	10.7	—	—
70-90	—	15.0	—	—

FUEL CONSUMPTION
Overall mpg: 14.6 (19.3 litres/100km)
Touring mpg*: 18.8mpg (15.0 litres/100km)
Govt tests mpg: 15.6mpg (urban) 26.9mpg (steady 56mph) 21.0mpg (steady 75mph)
Grade of fuel: Four star (97RM) unleaded (95RM) or (98+RM)
Tank capacity: 17.5 galls (80 litres)
Max range*: 329 miles
* Based on Government fuel economy figures: 50 per cent of urban cycle, 25 per cent each of 56/75mph consumptions.

BRAKING
Fade (from 75mph in neutral)
Pedal load (lb) for 0.5g stops

	start/end		start/end
1	27-30	**6**	30-45
2	30-35	**7**	30-40
3	30-35	**8**	30-40
4	30-40	**9**	30-40
5	30-40	**10**	30-40

Response (from 30mph in neutral)

Load	g	Distance
10lb	0.18	167ft
20lb	0.34	89ft
30lb	0.65	46ft
40lb	0.70	43ft
50lb	0.75	40ft

WEIGHT
Kerb 4379lb/1988kg
Distribution % F/R 50/50
Test 4835lb/2195kg
Max payload 1150lb/522kg
Max towing weight 8811lb/4000kg

TEST CONDITIONS

Wind	15-20mph
Temperature	7deg C (45deg F)
Barometer	992mbar
Surface	wet asphalt/concrete
Test distance	780 miles

Figures taken at 3100 miles by our own staff at the Lotus Group proving ground, Millbrook.

SPECIFICATION

ENGINE
Longitudinal, front, permanent four-wheel drive. **Capacity** 3947cc, 8 cylinders in 90deg V.
Bore 94mm, **stroke** 71.1mm.
Compression ratio 9.35 to 1.
Head/block al alloy/al alloy.
Valve gear ohv, 2 valves per cylinder.
Ignition and fuel Electronic ignition. Lucas L-Jetronic electronic fuel injection.
Max power 185bhp (PS-DIN) (138kW ISO) at 4750rpm. **Max torque** 235lb ft (319 Nm) at 2600rpm.

TRANSMISSION
4-speed automatic.

Gear	Ratio	mph/1000rpm
Top	0.728	26.9
3rd	1.000	19.6
2nd	1.480	13.2
1st	2.480	7.9

Final drive ratio 3.54 to 1. Transfer ratio high 1.206, low 3.244. Bevel centre differential with viscous coupling lock.

SUSPENSION
Front, live axle, radius arms, Panhard rod, coil springs, telescopic dampers.
Rear, live axle, radius arms, upper 'A' frame, coil springs, telescopic dampers.

STEERING
Recirculating ball, power assisted, 3.5 turns lock to lock.

BRAKES
Front 11.8ins (298mm) dia vent discs.
Rear 11.4ins (290mm) dia discs. Anti-lock standard.

WHEELS/TYRES
Cast alloy 7ins rims. 205 R16 Goodyear Wrangler tyres.

PRODUCED AND SOLD IN THE UK BY
Land Rover Ltd
Lode Lane, Solihull
W Midlands B92 8NW
Tel: 021-722 2424

COSTS

Prices

Total (in UK)	£31,949
Delivery, road tax, plates	£525
On the road price	£32,476
Options fitted to test car	None
Total as tested	£32,474

SERVICE
Major service 24,000 miles — service time 5.6 hrs. Intermediate service 12,000 miles — service time 4.15 hrs. Oil change 6000 miles — service time 2.70 hrs.

PARTS COST (inc VAT)

Oil filter	£11.59
Air filter	£6.61
Spark Plugs (set)	£13.62
Brake pads (2 wheels) front	£37.49
Brake pads (2 wheels) rear	£42.27
Exhaust complete	£339.25
Tyre — each (typical)	£68.77
Windscreen (heated)	£281.35
Headlamp unit	£77.94
Front wing	£172.50
Rear bumper	£99.73

WARRANTY
12 months/unlimited mileage, 12 months breakdown recovery

EQUIPMENT

Anti-lock brakes	●
Self levelling suspension	●
Alloy wheels	●
Auto gearbox	●
Power assisted steering	●
Electric mirrors	●
Steering rake/reach adjustment	—
Seat height adjustment	●
Electric seat adjustment	●
Lumbar adjustment	—
Catalytic converter	£450
Intermittent wipe (variable)	●
Heated windscreen	●
Leather trim	●
Air conditioning	●
Cruise control	—
Radio/cassette player	●
6 speakers	●
Electric windows F/R	●
Central locking	●
Front driving lamps	●
Headlamp wash	●
Electric tilt/slide sunroof	●
Metallic paint	●

● Standard — Not available

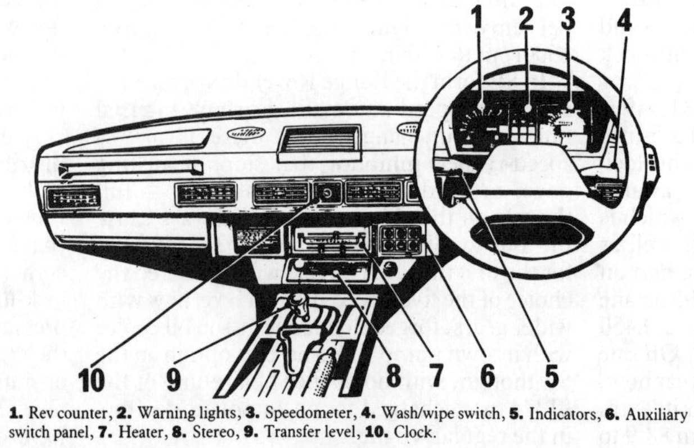

1. Rev counter, **2.** Warning lights, **3.** Speedometer, **4.** Wash/wipe switch, **5.** Indicators, **6.** Auxiliary switch panel, **7.** Heater, **8.** Stereo, **9.** Transfer level, **10.** Clock.

◀ rear wheels) and an exceptionally quick response time. The system samples the relative wheels speeds 250 times a second, and can cycle the brake pressure in each circuit up to 10 times a second to prevent lock-up.

None of which is apparent when you climb into the new SE for the first time. The instruments have gained new faces, making them slightly easier to read, but the slightly offset, thick-rimmed steering wheel and complex five-slider heating and air conditioning unit are unchanged. On the SE you sit in electrically-adjustable leather seats with new cushions that are as comfortable as your favourite armchair and deliver better support in all directions than before. They can provide huge amounts of legroom, but very tall drivers will still need to set the base as low as it will go to clear the electric tilt/slide sunroof.

Start the engine, though, and you know immediately something has changed. Gone is the deep-throated but benign V8 burble, replaced by a harder, more urgent engine note that frankly doesn't sound as good. Into 'D', hit the throttle and the better torque spread of the new engine makes itself felt. This engine pulls harder from much lower revs than its predecessor, with no loss of smoothness, though it does start to sound a little harsh toward the 4750rpm peak power speed. Despite a complete absence of electronic control, the ZF 'box slurs through the changes almost imperceptibly. Full or part throttle, up or down the ratios, it's every bit as good as the best autos from Mercedes or GM.

Flat out, the SE will now reach 108mph, 7mph faster than the old SE, at 4000rpm with the engine spinning over well below its peak power revs and sounding suitably unhurried. As usual there is no red line or rev limiter, but on full throttle the 'box changes up at around the 4800rpm mark. Held in each of the first three gears, it can reach 43, 73 and 108mph at a still-safe 5500rpm. From a standstill, the 3.9 is fractionally but noticeably quicker, reaching 60mph in 11.3secs and the quarter-mile mark in 18.5secs. That's 0.4secs faster than the 3.5-litre SE over both increments.

The extra torque also makes the 3.9 quicker through the gears, but more impressive is the punch available at lower revs. Now the SE can be hustled along a winding lane left in 'D'. In the old car, if you didn't want to use full throttle to activate the kickdown, the only way to ensure brisk progress was to lock out top gear by leaving the gearbox in '3'. There's a price, of course; the 3.9 Vogue returned just 14.6mpg overall, but that is only just over 1mpg worse than the 3.5-litre Vogue SE we tested.

There's a completely different feel to the brake pedal with the Wabco anti-lock; it's firmer and there's less initial travel but more progression and bite. In the wet the anti-lock generates a rapid pulsing action at the pedal that brings the SE to a quick stop in a straight line, and our repeated tests showed rather better resistance to fade despite the near-200lb hike in kerb weight to 4379lb.

Wind noise is still not one of the Range Rover's strong points, with the tops of the door frames the main culprit above 90mph. There's more soundproofing and barely any engine or road noise find their way into the cabin.

With the familiar long-travel coil springing, high centre of gravity and lack of anti-roll bars, the Range Rover rolls, as always, to what feels like a crazy angle under hard cornering. But the progressive understeer provided by permanent four-wheel drive and the perfect camber control of live axles make it a secure and surprisingly capable companion. Despite the high unsprung weight delivering a touch of harshness over very small bumps, that same suspension provides a supple ride of luxury saloon quality on the road. The rougher the road, the more the Range Rover pampers.

Electric leather seats boast new cushions that improve side and lumbar support. Headroom still limited by electric sunroof. New instruments have clearer graphics but still no rev counter red-line

The power-assisted recirculating ball steering still conspires with the steering damper to prevent kickback off-road but gives a lifeless, unresponsive feel at speed. A change to Goodyear Wrangler dual-purpose tyres, of the same 205 R16 size as the Michelins they replace, has given marginally more precision in a straight line at motorway speeds.

Off-road, the Range Rover's seemingly endless trek upmarket has not compromised its ability one iota. The chassis still delivers a staggering level of grip in the stickiest going imaginable, with a level of axle articulation, ground clearance and ride comfort that puts its rivals to shame. The extra urge of the 3.9 and its response from lower speeds help the Range Rover tackle some climbs in a higher gear than before, which makes it easier to keep the wheels turning.

Rule one of off-roading is don't touch the brakes. It's still rule one, but Land Rover recognises the attractions of the left-hand pedal to some owners in a panic situation and has made sure that the anti-lock system is equally effective off the beaten track. You can brake when descending a slippery slope with no danger of the wheels locking and losing control and, in fact, a gentle squeeze on the pedal can compensate for the auto's lack of a really low first gear.

Every time the Range Rover's price tag passes a landmark, be it £20,000, £25,000 or £30,000, the value for money of a 20-year-old design is questioned. There's no doubt Land Rover makes a very healthy profit on the Vogue SE, but the company points out that it is in the same market and price range as a Jaguar Sovereign or BMW 735i.

The SE is as well-equipped, roomy and comfortable as most luxury saloons, offers perfectly adequate road performance and is still the only car that combines all that with superb off-road performance. With Discovery to tackle the lower-priced, recreational end of the market, the Range Rover doesn't need to be cheap, and it isn't. But for those who recognise the Range Rover's unique abilities, the new Vogue SE is worth every penny. ■

John Beese tests 4.5 litres of RoverCraft muscle

power politics

What does £8000 buy you these days? Well, if you own a good condition Range Rover V8 Efi and would like to dramatically increase its all-round performance, this is the sort of smelly stuff you'll need to secure: Maidstone-based RoverCraft engineering's all-singing all-dancing 4.5-litre, MoTec fuel-injected package including revitalised suspension and brakes. And we are talking of a serious power increase here. If you're selfish enough to use leaded petrol RoverCraft claim to take the V8 3.5-litre Efi's power from 165bhp up to 300bhp and beef up its torque from 207lbf.ft to a stumping 320lbf.ft. Because the Range Rover is such a heavy car it's the increased torque that improves acceleration and helps with heavy towing. There are many relatively simple modifications you can make to the V8 to increase the brake horse power but to alter

●**RoverCraft's 4.5-litre conversion takes the 3.5-litre Efi engine's performance from 165bhp up to a claimed 300bhp**

the torque characteristics seriously you're looking at a complete engine rebuild.

However, improving the power of your Range Rover needn't mean mega bucks. RoverCraft offer an enormous variety of performance packages for the V8 engine in fuel-injected or naturally-aspirated form, which are available as a completed package or in kit form. For example, the part exchange price on RoverCraft's performance block assembly is £710. This includes a reground and Tufftrided crankshaft, heavy-duty main and big end bearings, new camshaft bearings, rebored cylinder block fitted with high compression pistons, balanced crankshaft assembly and refaced cylinder block. In kit form (part exchange) this level of conversion would set you back around £590. The company also offer a complete machining and tuning service.

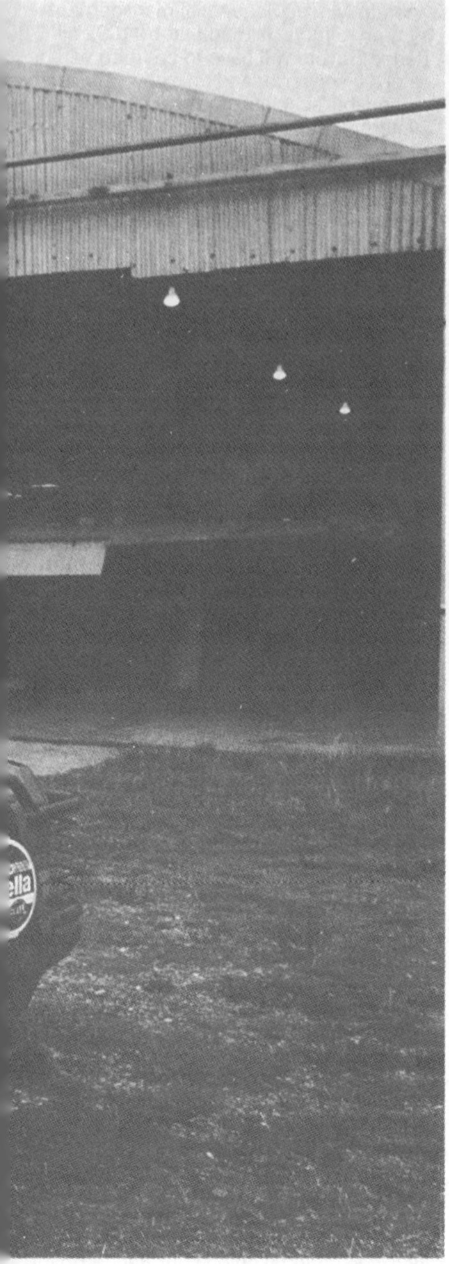

The F-registered automatic gearbox V8-powered Range Rover we had on test had been bored out to 4.5 litres and given the full RoverCraft treatment (except for the brake package). Apparently the vehicle is frequently used to tow a heavy boat long distances and the customer was looking for a large increase in mid-range power. We weren't able to tow a heavy trailer to test the vehicle's pulling power, but its overall performance was spectacular, with the revitalised engine delivering its clout with an evenness only obtainable from a sophisticated engine management system. In this case RoverCraft had opted for the Australian-manufactured MoTec Sequential unit which accounts for £1000 of the £8000 conversion.

Mr RoverCraft is Simon Law, who explained that before such a conversion can begin the vehicle needs a thorough examination. Clearly there's little point in investing this kind of money in a car with worn mechanics. Most customers who opt for the full treatment have Range Rovers that are at most only a few years old. The latest specification Borg Warner transfer box with viscous-operated locking centre differential is, according to Simon, superb at coping with the extra power.

After the 3.5-litre unit has been stripped down and checked for wear, all the parts are chemically cleaned and the cylinder heads bead blasted. The cylinder block is then machined to take the new liners and bored to match the forged pistons (93.5mm). RoverCraft fit a long-stroke crank of their own design, offering a stroke of 82mm against the standard 71.1mm, and to cope with the increased stress on the main bearings they have uprated these from 2.3cm to 2.5cm diameter and had them manufactured from 700 nodule cast iron in place of the current 600 specification.

According to RoverCraft there are no problems with boring out the 3.5-litre engine to this extent, but the life and performance of the engine depend on how well the entire conversion is designed and constructed. The flywheel is re-balanced and lightened to improve performance, and a heavy-duty clutch fitted. The new crank is Tufftrided (the outer case is hardened) and balanced. The conversion also benefits from RC87 camshafts with hydraulic followers and the rocker shafts are replaced with steel items. The cylinder heads are fitted with triple-angle valve seats and Vitesse inlet and exhaust valves are used. All the valve intakes are gas-flowed to test efficiency and modified where necessary. The valve guides are lowered and reprofiled, and fitting measurements are checked when the engine is assembled. The cylinder heads are then refaced to achieve the

● **The power-house Range Rover flexing its muscles at two 15-ton Alvis Salamander 6×6 fire tenders – capable of producing 90,000 gallons of foam in two minutes!**

power politics

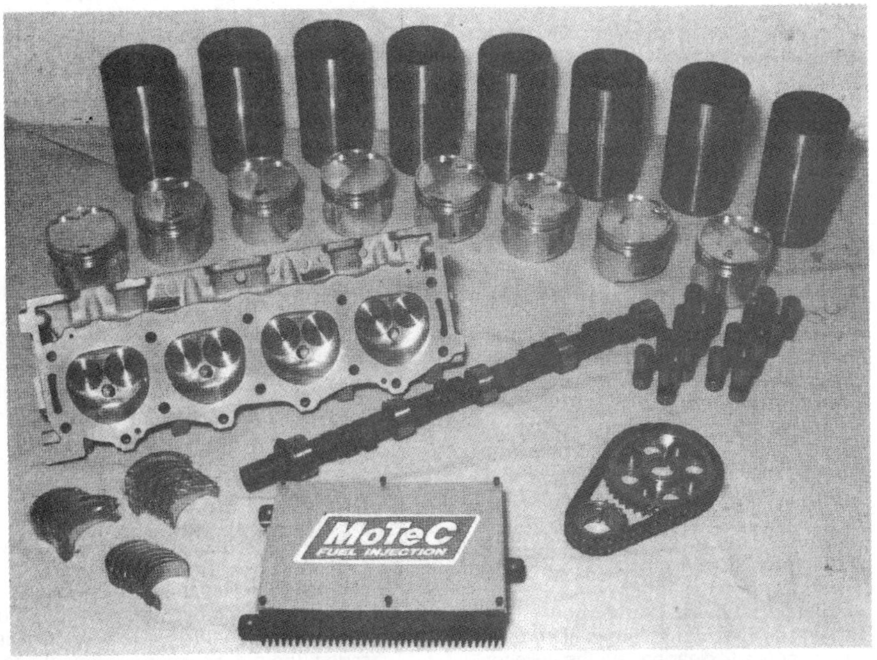

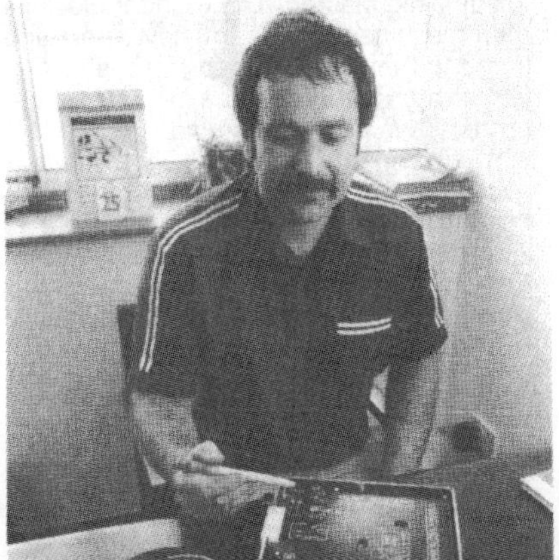

● **Above: new liners, high compression pistons, reground cylinder head, RC87 camshaft, hydraulic followers, uprated main bearings and timing gear plus the MoTec fuel injection control box – all part of the conversion. Right: Simon Law explaining how to programme the MoTec system. Below: the RoverCraft-designed crankshaft after casting and (below) after Tufftriding and machining**

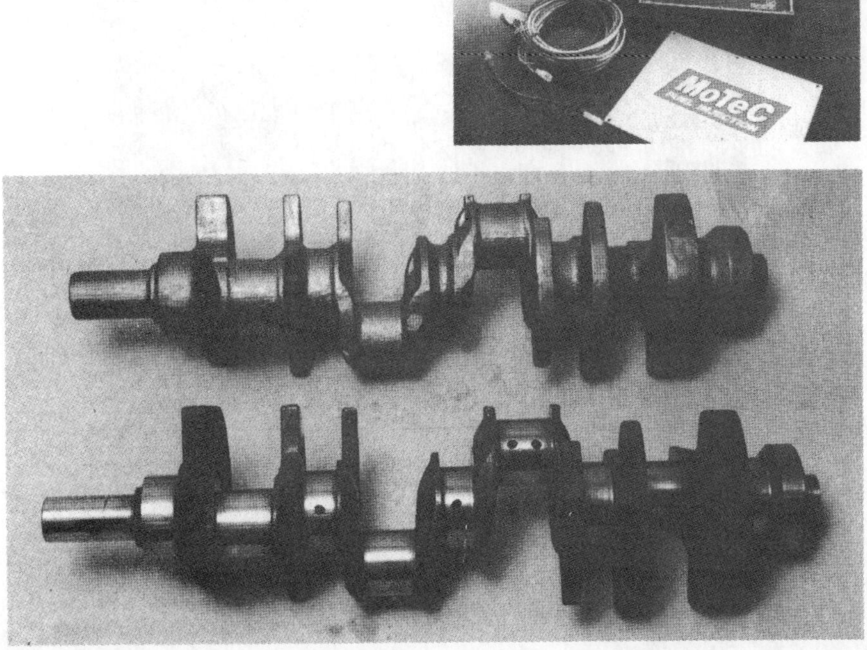

correct compression ratio and a new modified oil pump fitted.

If you just opt to have this work carried out on the engine without the MoTec engine management system or uprated suspension and brakes, the bill will be around £3600, or £1818 in kit form (part exchange). Clearly, if you are able to undertake the work yourself you can save a considerable amount of money.

RoverCraft point out that, however good the engineering is on a conversion, it still depends on a well-controlled fuel and air supply to run efficiently. Simon Law told us that some fuel-injected conversions suffer from poor quality systems, and this is why RoverCraft recommend the relatively expensive but efficient and simple-to-operate MoTec Sequential management system. Simon claimed that by 1992 all Range Rovers will have to be fitted with a catalytic converter and unless you have a reasonably sophisticated management system, they will not be able to cope with the increased demand variations across the power band caused by the converter.

The MoTec system in the vehicle we tested was neatly located under the driver's seat. Staring at the smart green electrical box of tricks, measuring about 10 inches square and two inches deep, it was difficult to believe that it is capable of monitoring the fuel supply at all the varying load characteristics displayed by such an engine, minutely altering the rate and type of supply in fractions of a second in addition to monitoring temperature changes, water temperature and inlet temperature variations and even altitude differences.

Once the unit is wired up you need to plot the volumetric efficiency of the engine at various load ratings by running the vehicle on a rolling road and keying the metered data into the memory of the MoTec unit. The intention when setting the metering is to get the fuelling to follow what the torque curve is doing. The theory is that once programmed you only ever need to adjust the idle mixture and the MoTec system will accommodate all the aforementioned variations.

The handling package fitted to the vehicle we tested consisted of uprated springs and a conventional anti-roll bar set-up. With the significant improvement in acceleration any modifications that help keep the well-known Range Rover roll in check are welcome. Although the package limits the ultimate off-road capability of the vehicle by reducing axle articulation, you can rest assured that this vehicle would fulfil most people's off-road desires when driven correctly. RoverCraft's brake modification consists of fitting larger capacity calipers and is particularly recommended in situations where constant hard braking is needed – such as driving in Alpine conditions.

RoverCraft's conversion is clearly an upmarket package aimed at those who don't have to compromise on their choice of transport. If you do have the resources to spend and want a high-quality performance package undertaken on your Range Rover then this is well worth considering. For more details phone 0622 687070.

PHOTO BY GUNNAR CONRAD

RANGE ROVER COUNTY

Conquering the Continental Divide

BY JOHN DINKEL

"SHOCK. IT'S LIKE being in shock," said Charles Hughes, president of Range Rover of North America, Inc. "A quasi-religious experience," piped in Bill Baker, Range Rover's vice president of corporate communications.

The "experience" they were referring to was simply the beholding of breathtaking views in the Colorado Rockies that our expedition had come upon. It was impossible to round a bend or top a rise without coming across what everyone knew had to be the most spectacular scenery he had ever seen.

Range Rover is a small company with big ideas. Consider for a moment: A Range Rover isn't a conventional automobile. So, logic would dictate an unconventional preview for an unconventional vehicle. How best to show off the assets of the 1990 Range Rover with its recontoured leather seats, its revised instrument lighting and graphics, its improved heater/vent/air-conditioning system and, especially, its sophisticated anti-lock braking system (ABS)? How about staging the Great Divide Expedition, the first-ever continuous journey by motor vehicle to follow the Continental Divide?

Why Colorado? Because this route just happens to be the most difficult border-to-border route of any state in the continental United States and the highest continuous road in North America, traversing elevations from 7000 to more than 13,000 feet above sea level. For two weeks the Expedition would cross and parallel the Great Divide for approximately 1000 miles (less than 300 on pavement), traveling through the state of Colorado from the starting point at Encampment, Wyoming to the finish at Chama, New Mexico. It would wind through 32 mountain passes, 15 of which cross the Divide and only four of which are paved.

Aspen resident Tom Collins, a conservationist and a Range Rover Camel Trophy co-driver, was the route architect. His route took us over mostly unpaved roads and 4-

wheel-drive trails originally carved out by Indians and early miners. We drove through seven national forests, past historic ghost towns and more abandoned mines than you could shake a stick of dynamite at.

Aside from stressing the Range Rover's durability, reliability and off-road sure-footedness, the British carmaker used this event to promote the U.S. Forest Service's Tread Lightly program, a national educational effort to encourage environmentally responsible and safe off-highway driving techniques.

Assisting in the Expedition was world-famous off-road racer Malcolm Smith, who has been a Range Rover fan since driving a specially built prototype to the highest finish ever (4th) for an American in the grueling 8000-mile, 22-day Paris-Dakar rally in 1988. He is also a six-time winner (three times on a motorcycle and three in a car) of the Baja 1000, a four-time winner of the Baja 500; and he has twice won the Mint 400 and the Roof of Africa Rally.

The event was staged in three waves and included journalists from as far away as England and Australia. I hooked up with the Expedition as part of Group III the evening of the ninth day in Ouray, named in honor of the Ute Indian chief who kept the peace between whites and Indians.

With nine days in the books, all had gone extremely well, save for one "moment" on the morning of the first day when brain-fade had resulted in the loss of one Range Rover to a rollover. Thankfully no one was hurt, and the expedition continued minus one vehicle.

Next day we were up by 5:30 a.m. and on the road by 7:00, a ritual we would follow for the entire event. We had 61 miles to travel and we'd need at least seven hours to cover them, so we couldn't tarry. Our route took us over fabled Black Bear Pass (12,840 ft.), which is a must on every serious off-roader's list of places to drive. But it's no place for anyone with acrophobia, what with its myriad drop-offs and switchbacks.

We had numerous opportunities to test the Range Rover's ABS while driving over this pass. This system took four years to develop because Range Rover wouldn't accept an off-the-shelf system that would have required performance compromises with the vehicle's 4-wheel drive. I give the electronically controlled 4-channel ABS high marks for directional stability and controllability under the toughest of off-road conditions. The system is fully compati-

Subtle but significant detail changes mark the 1990 Range Rover interior. Seats are recontoured, instrument lighting and graphics are revised and the heater/vent/air-conditioning system is much more effective, especially in the flow-through, non-a/c mode.

The Range Rover's beauty is more than skin deep. New this year is an anti-lock braking system that was four years in the making. The electronically controlled 4-channel ABS is fully compatible with all aspects of the Range Rover's sophisticated 4-wheel drive.

PHOTOS BY RON PERRY

ble with full-time 4wd and is unaffected by the viscous coupling unit in the center differential or the selection of high or low range on the transfer case.

At the end of each day the Range Rover mechanics would knock the dust out of every air filter, check the Michelin tires (the identical Michelins that come on U.S. Range Rovers) and fill the fuel tanks. Oh, one more thing. Any driver who dinged or scratched a vehicle was required to sign his artwork with a felt-tip pen.

Each of my four days was better than the last. More than a few books have been written on the personalities (Calamity Jane, Doc Holliday, Kit Carson, Wyatt Earp, Bat Masterson and Butch Cassidy, to name a few), the horrific struggles of those early settlers and the incredible natural beauty of the Colorado Rockies. Four days will never be enough; I know I'll be going back.

We popped the cork from a bottle of champagne at the New Mexico border as a symbolic ending to our journey. I left with an even greater appreciation of the competency of the Range Rover on and off road.

After the expedition, we had the opportunity to track test the newest Range Rover, the luxurious County edition, at our California home base. Last year, the Range Rover acquired the 3.9-liter edition of the tried-and-true 3.5-liter aluminum V-8 that Rover bought from Buick. Horsepower is now 178 at 4750 rpm (versus 150 bhp in the last Range Rover we tested—November 1987) and torque measures 220 lb.-ft. compared with the earlier 195. All in all, the Range Rover just keeps getting better and continues to set new standards in the luxury 4wd category.

Test Notes . . .

■ Despite its great weight, the Range Rover's new anti-lock brakes allow respectable stopping distances, though control is less impressive than we've come to expect with ABS-equipped vehicles.

■ Maximum cornering grip is modest at 0.65g, though the Range Rover always feels secure and stable despite its body roll angle, which can become significant.

RANGE ROVER COUNTY

0–60 mph .... 12.3 sec
0–¼ mi .... 18.7 sec
Top speed .... est 109 mph
Skidpad 0.65g
Slalom .... na
Brake rating .... good

PRICE

List price, all POE .... **$40,125** Price as tested **$40,125**

Price as tested includes std equip. (air cond, ABS, AM/FM stereo/cassette/CD, leather interior, elect. window lifts, elect. adj seats, elect. adj mirrors, cruise control, central locking).

PHOTO BY RON PERRY

ENGINE

Type	ohv **V-8**
Displacement	241 cu in./3947 cc
Bore x stroke	3.70 x 2.80 in./ 94.0 x 71.1 mm
Compression ratio	8.1:1
Horsepower (SAE):	**178 bhp @ 4750 rpm**
Torque	**220 lb-ft @ 3250 rpm**
Maximum engine speed	5250 rpm
Fuel injection	Lucas elect. port
Fuel	prem unleaded, 91 pump oct

GENERAL DATA

Curb weight	**4485 lb**
Test weight	**4600 lb**
Weight dist, f/r, %	**52/48**
Wheelbase	100.0 in.
Track, f/r	58.5 in./58.5 in.
Length	175.0 in.
Width	71.4 in.
Height	70.8 in.
Trunk space	13.0 + 6.8 cu ft

DRIVETRAIN

Transmission .... **4-sp automatic** with lockup torque converter

Gear	Ratio	Overall ratio, high range	(Rpm) Mph
1st	2.48:1	10.59:1	41
2nd	1.48:1	6.32:1	69
3rd	1.00:1	4.27:1	est 102
4th	0.73:1	3.11:1	est (4120) 109

Transfer ratios; high/low	1.21:1/3:24:1
Final drive ratio	3.54:1
Engine rpm @ 60 mph in 4th	2270

CHASSIS & BODY

Layout	**front engine/4wd**
Body/frame:	aluminum & steel/steel ladder
Brakes, f/r	**11.8-in. vented discs/ 11.4-in. discs;** vacuum assist, ABS
Wheels	cast alloy, **16 x 7**
Tires	Michelin X M+S 244, **205R-16**
Steering	**worm & roller,** power assist
Turns, lock to lock	3.4

Suspension, f/r: **beam axle** on Panhard rod, leading links, coil springs, tube shocks/**live axle** on trailing links, upper A-arms, coil springs, tube shocks

ACCELERATION

Time to speed	Seconds
0–30 mph	3.9
0–60 mph	12.3
0–80 mph	22.0
Time to distance	
0–100 ft	3.8
0–500 ft	10.1
0–1320 ft (¼ mi)	18.7 @ 74.0 mph

BRAKING

Minimum stopping distance	
From 60 mph	153 ft
From 80 mph	270 ft
Control	good
Pedal effort for 0.5g stop	20 lb
Fade, effort after six 0.5g stops from 60 mph	22 lb
Brake feel	fair
Overall brake rating	good

FUEL ECONOMY

Normal driving	13.5 mpg
EPA city/highway	13/16 mpg
Fuel capacity	20.0 gal.

HANDLING

Lateral accel (200-ft skidpad)	0.65g
Balance	moderate understeer
Speed thru 700-ft slalom	na
Balance	na

INTERIOR NOISE

Idle in neutral	51 dBA
Constant 70 mph	72 dBA

Subjective ratings consist of excellent, very good, good, average, poor; na means information is not available.

Howard Lees takes a latest spec Range Rover Vogue to the Arctic Circle and finds ice driving an unforgettable experience

LAP TIMES

When it came, the crack was like a rifle shot, splitting the silence around the frozen lake and making me sprint the 50 yards back across the ice to the Range Rover as if my life depended on it. I had a great deal of respect for the Turbo D's wading ability, but I wasn't keen to test its prowess as a submarine.

Any Finn would have read the warning signs. We were a good 400 miles south of the Arctic Circle now, and the temperature had been hovering around zero for a week. The lack of snow on the branches of the pines lining the shore was another clue – even just a few yards from the shore, the ice simply wasn't thick enough to support a Range Rover for long.

As the saying goes, it had seemed like a good idea at the time. Looking for a good test of the new roll bar-equipped Range Rover over Christmas, the Arctic Circle sounded like an appropriate destination in the middle of winter. Rovaniemi in Northern Finland has the required 66 degrees, 33 minutes northerly latitude, although it's only on the southern fringes of Lapland. The plan was to arrive at the Arctic Circle on December 21, the shortest day of the year, when the night would last the full 24 hours.

Winter tyres are a legal requirement in Finland anyway during the winter months, but to cope with the permanent ice cover that far north we wanted to run studs as well. The Range Rover's 205R16 Michelin XM+S244 tyres fit the former requirement, and also have the necessary holes ready for studs. We didn't want to carry two sets of wheels, and running on studs in England is legal, so Greyhound Tyres in Shepherds Bush fitted the studs with just a few hours' notice. Studs are definitely not legal in Germany, so our route was decided for us – ferry from Harwich to Goteborg, then across Sweden for the boat to Finland.

Since I was paying for the fuel, I'd already settled on a Vogue Turbo D – and I was curious to see how the Scandinavians kept their diesel engines running in temperatures down to minus 40 degrees C or so. Our other preparations were straightforward – packing snow chains just in case, a good shovel,

● *Standing proud on studded Michelin XM+Ss*

towrope, toolkit, some spares and a decent board for under the jack.

Tyre studs need running in just like tyres, so we were careful to keep the speed down to 65mph or so on the way to Harwich. They played a constant tune in the background, like surf on a distant shore, though the grumble from the 2.5-litre VM turbo-diesel was doing its best to mask that. With just 1900 miles on the clock, this particular engine was a long way from being the smoothest Turbo D unit I have sampled – the 2.5-litre unit is a shade lumpier than its less flexible 2.4-litre predecessor, but I hoped it would smooth out with a few more miles on the clock.

Rumbling down the ramp off Goteborg's Skandiahamnen terminal at 5pm we were faced with a 300-mile trek across Sweden in the dark, not ideal circumstances for developing a feel for ice driving. In fact, though, the roads were clear of snow and heavy salting seemed to have kept the ice at bay.

Half-way through the trip it started to snow heavily. The road surface glistened in the light of the headlamps as I tried to decide whether it was ice or not. Sometimes it was, sometimes not – I found I could get a feel for the available grip through the steering wheel. It was midnight before I cut the engine outside our Stockholm hotel. Next morning we crossed to Turku and then picked our way north-east, our destination that night being Lahti, some 90 miles north of Helsinki.

Working the Turbo D harder now, it was clear the anti-roll bars had improved its poise and control through a series of bends – less roll and better steering response. Sometimes,

though, the combined effect of the tyre studs and the several inch-deep ruts created in Finnish roads by their almost universal use meant the Range Rover could feel a little loose at speed. The temperature was well below zero out of town, but still the roads were wet, salt combining with flurries of snow to produce a dirty, tenacious brown coating. 30 miles outside Lahti the studs started to work hard for their living. Snow was blanketing down, cutting our forward view to a few yards and coating the roads to a depth of several inches – again, there was sheet ice underneath.

It was colder still the morning we left Lahti. There was no spray now – just a satin-smooth coating of ice on the roads with flurries of snow blowing across it. By 9am it was barely light enough to see without headlamps.

We were grateful of course for the Range Rover's permanent four-wheel drive and Wabco ABS, but also for the strong torque the VM engine produces as low as 2000rpm. With another 1000 miles on the clock it was already starting to smooth out. On the single carriageway roads that make up the overwhelming majority of Finland's infrastructure we made decent progress.

By the time we reached Jyvaskyla – home of the 1000 Lakes rally – there was at least 18 inches of snow cover in the fields, and the banks left at the sides of the roads by the snow ploughs were twice as high. I was starting to explore the limits of the Range Rover's grip now, and noticed that the noise the studs made when braking or accelerating gave another clue to the amount of grip available. You could hear impending lock-up, the muted expresso-machine hiss of the 108 hardened steel spikes per tyre deepening just before the anti-lock brakes started to pulse.

About 80 miles north of Jyvaskyla we came across our first official ice road, across the lake at Viitasaari. With 60,000 lakes scattered throughout the country, the occasional short cut across the ice can save dozens of miles, and some are ploughed into roads by the local authority. This one had a sign giving a two-tonne weight limit, but though we weighed closer 2.5 tonnes we reckoned they'd probably allowed a safety margin, and headed down the ramp onto the ice.

It was a strange feeling, driving past boats locked into the ice for the duration, but as confidence built I soon found the Range Rover was in its element. It had terrific traction on the ice, but in high range first it could be powered round in a series of glorious oversteering drifts. Off the ploughed track there was about a foot of snow, but the Turbo D didn't notice this at all and barely slackened pace.

We spent the night at Oulu; Rovaniemi was barely 150 miles north. On a wide, thick ice road the next day there was virtually no traffic,

● *The highlight of the trip was sampling four-wheel drive on a frozen lake*

The turbo-diesel unit returned an average of 26mpg

Arriving at Napapiiri and entering the Arctic Circle

LAP TIMES

The armchair comfort of the Range Rover reduces fatigue on long journeys

but even with the chance to run on main beam all the time I missed the spoiler-mounted driving lights of the V8 Vogue. The ice here looked almost black and had black hole-like ability to leach away what little light the Range Rover's headlamps could project.

The Turbo D had covered a full 430 miles since its last fill when we stopped for fuel: an impressive range even given the larger plastic fuel tank – over the course of our 2500-mile round trip the Range Rover averaged 26mpg. We also discovered how even the large logging trucks run all year round – Shell's winter diesel is guaranteed not to wax up above minus 38 degrees C, and there was also a slightly more expensive arctic grade on sale good for minus 43 degrees. In fact the temperatures were unseasonably mild, and even at night didn't sink below minus 15 degrees.

The characteristic clusters of tents on secluded lakes confirmed that this was Lapland. Five miles north of the town, the Arctic Circle bisects the main road north at Napapiiri – it's also the site of Santa's workshop, so we headed away from the tourists in search of some logging trails. Clearly these aren't used much in the winter, so they provided our opportunity to flex the Range Rover's off-road muscles in deeper snow. The problem was that the snow smoothed out the contours, and unlike on a lake, you simply couldn't tell how deep it was underfoot without walking every yard. In low range with the viscous centre diff lock working overtime, it wasn't a lack of traction that defeated the Turbo D – rather it came to a stop when it had bulldozed such a large bow wave of snow ahead of it that it simply couldn't push any more. We could probably have gained a few more yards by breaking out the snow chains, but we'd learnt enough and retraced our steps.

A 6am start for the 500-mile trip back the next morning meant the majority of our Arctic Circle wandering was destined to be in darkness, and we came to realise why most of the local cars were festooned with more driving lamps than the average rally car.

20 miles south of the town, a veritable Christmas tree of flashing orange lights ahead turned out to signal the sight of a little local indiscretion. The Ford Cortina well and truly stuck 20 feet down the steep snow bank at the side of the road was about the 10th impromptu snowplough we'd seen so far, but this one was just minutes old. No-one was hurt, but none of the assorted tracks, vans and cars that stopped had managed to haul the car out. After almost falling on my face on the sheet ice of the road surface, I was not confident the Range Rover would do the job either, but it was clear all the locals expected me to try.

Tow rope hitched up, I aimed at about 45 degrees to the road, selected low range and tried a snatch start. The engine stalled, showing more grip than I'd expected, so I tried again with more revs. This time the studs bit deep into the ice and the Cortina jerked forward a couple of feet. Three more attempts and it was back on the road, on its wheels and still a runner – by the effusiveness of the owner's thanks he had clearly expected to be rather later for work that morning.

Our trip back ran like clockwork, the Turbo D still running 400 miles on each 18-gallon tankful and getting smoother with every mile. Even with no more than a few hours of daylight every 24 hours, covering miles in the Vogue Turbo D was no hardship – all I would ask for would be better lights, headlamp wipers and an outside temperature gauge. In a week I learnt a lot about driving on snow and ice – that there are as many different surfaces as there are types of sand in a desert, and the parallels go further. Just like driving on sand, the secret is reading the surface, looking ahead at colours and textures, feeling for grip through the steering wheel and, a new twist, listening for a change of note from the ice studs as they bite into the ice. But the best bit of the trip was sampling four-wheel drive on a frozen lake – if that was all northern Finland had to offer, it would be enough to make the trip worthwhile.

RANGE ROVER CSK

Range Rover is only building 200 CSKs. A lot of people could be disappointed, suggests Simon Arron

THE SCENE is a public house, anywhere in England. The jukebox clicks into action, and the voices struggle to compete with another 50 pence worth of Dire Straits. Above the monotonal guitar, snatches of conversation can be heard. . .

"Blah, blah, mutter . . . XR3, blah, natter, GTi, blah, blah, turbo . . . blah, 16 valves, blah, blah, XR2, blah . . . blah, 205, mumble, mumble, Cosworth . . . grunt, P600s, blah, blah, CSK, blah, blah . . . *you what?*"

"Er, CSK."

"Mumble, grunt, CSK?"

Range Rover would, one presumes, never be so vulgar as to label one of its products as a GTi, but that is more or less what the limited edition CSK amounts to. The acronym is a tribute to Spen King (full title Charles Spencer King), original designer of the Range Rover concept which – 20 years after its launch – has yet to be bettered. There have been many aspirants to the Range Rover's throne, but a gradual process of refinement has kept it ahead of its (mainly Japanese) imitators over the years.

In most market sectors, the Oriental policy of replacing models every three of four years pays off. Here, however, the Range Rover has hardly changed (at least to the naked eye) after two decades, and still it rules the roost.

At the heart of the CSK is the familiar, fuel-injected Rover V8, in 'stretched' 3.9-litre guise. As it does in the Vogue SE, this thumps out 185bhp at 4,750rpm and 235lb ft of torque at just 2,600, but the lighter three-door shell and mildly revised aerodynamics make this the fastest ever production Range Rover. Claimed top speed is 114mph (the Vogue is only around 7mph slower, but it's difficult to make too many gains in this area, given that Range Rovers are about as aerodynamic as Finland). Accelerative improvements are more significant. The five-door Vogue SE lumbers from 0-60mph in a respectable 11.3sec; in the CSK, Range Rover claims that this is reduced to just 9.9. For a vehicle of this size and weight, that's going some.

It *feels* quick, too, leastways until air molecules start putting up stiff resistance at around 85mph. Around town, it is as responsive as any sporting hatch you might care to mention, although its size dictates that you could never really label it as 'nippy'. Perhaps the CSK could be the forerunner for a whole new breed, although there is no obviously catchy handle in the 'hot hatch' mode. Hot barge, maybe?

Not only will the CSK outrun previous Range Rovers, it will also outcorner them. Although tenacious, the Range Rover's way has always been to teeter around bends with all four wheels at eccentric angles. Such untidiness *can* still be achieved with the CSK, but it now requires greater effort following the adoption of anti-roll bars and sports damping. This has increased roll stiffness by 25percent, the benefits of which are immediately apparent. Creditably, this has been achieved without spoiling the ride, although the reduced suspension articulation over a given bump will inevitably compromise its off-road ability to an extent.

The CSK is available as either a five-speed manual (as tested) or four-speed auto, and standard touches include liberal interior splashes of

STEVE BURKE

leather and walnut, anti-locking braking, air conditioning, electric everything, front fog lamps and a six-speaker sound system.

Add the gorgeous V8 burble and a price tag around £3,000 shy of the Vogue SE's (in manual guise – the auto is £1,400 extra), and you have a simply superb all-rounder which is genuinely enjoyable to use. As we went to press, there were still some available. Buy one while you can.

Price: £23,995 (manual); £30,319 (auto). Verdict: In a word, brilliant.

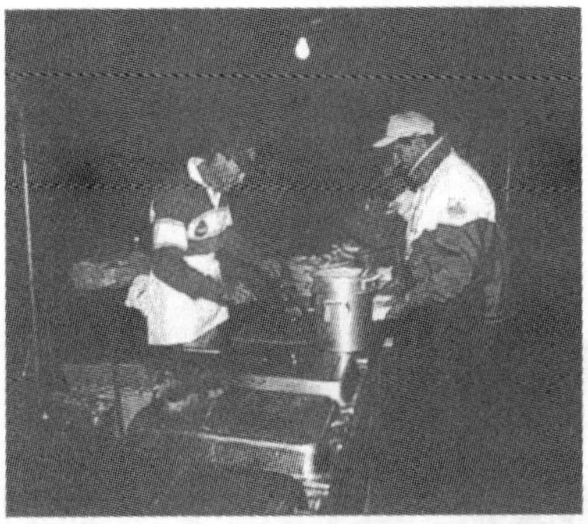

The Americans have a reputation for never doing things by halves. And their handling of the Range Rover is no exception, as Howard Lees reports

There are more 4x4s sold and probably more off-roading done in the USA than anywhere else in the world – but only since the Range Rover arrived four years ago have US buyers been able to get hold of a really versatile vehicle. But with awareness of the Land Rover and Range Rover names, let alone the capabilities of the vehicles, at a very low level, Range Rover of North America had to do something to spread the word.

First off was the Great Divide expedition in 1989, a 1100-mile journey with eight Range Rovers that marked the first continuous north-south crossing of Colorado's 14,000ft high Continental Divide by motor vehicle. That got plenty of publicity, but didn't give Range Rover owners first-hand experience or tutition in what the vehicles could do. What was needed was an off-road driving school, but with US Range Rover owners used to the best and not short of a bob or two, a mere driving school wasn't enough – RRNA's version is the £2000-a-place Range Rover Driving Academy.

The man charged with setting up the Academy was Tom Collins – organiser of the Great Divide expedition and a regular Camel Trophy man. Collins and RRNA were both keen to build on the success of the Great Divide expedition, yet needed a base that could provide the sort of comforts customers were used to. The ski resort of Aspen fitted the bill, its 8000ft location giving access to passes up to 14,000ft within a day's drive and allowing the use of the five-star Jerome Hotel as a base.

Having luxury accommodation was the key to attracting the right kind of customer, but it didn't mean the off-roading would be child's play. Collins used some sections of the Great Divide route, meaning the highest passes would only be free of snow and passable from mid-August to mid-September. A Range Rover mailshot to 12,000 potential customers brought 100 firm deposits by return.

The vehicles were limited edition 'Great Divide' Range Rovers – US-spec 3.9-litre Vogue automatics but with Discovery-style alloy wheels and a front bull bar rather than the usual spoiler. Five were allocated to the students and one as the lead vehicle, and there were five instructors to back up Tom Collins, including Solihull's demonstration guru Don Green. Day one was spent in the classroom and at a local racetrack learning the basics. Steep climbs and descents, side slopes, basic winching, slalom tests and ABS demonstrations on gravel left many customers wide-eyed.

Monday was spent driving up the rear of Aspen Mountain, over the 12,000ft Taylor Pass. The descent included some particularly nasty, narrow, rocky tracks where the Range Rovers had to be guided down on foot, but all survived unscathed. The next day the group went further afield to the 19th-century coking town of Redstone, then over the 10,800ft Schofield Pass. Tom Collins described this as 'much trickier', with large rocks and steep descents to build people's confidence. With more faith in their Range Rovers, the party then tackled the muddy climbs and steep, off-camber wet rocks of Pearl Pass before heading back to the Hotel Jerome.

A day's rest from off-roading gave everyone a chance to sample a whole range of other activities – horse-riding, hot air ballooning, white water rafting and so on – all part of the package. Thursday marked the start of a two-day trek that would include a night under canvas, but as Tom Collins pointed out: 'Range Rover-style camping is not roughing it.' An encampment had already been set up in the 10,000ft meadows of Taylor Park, complete with cordon bleu catering and a not inconsiderable supply of alcohol.

En route to the overnight halt, the six Range Rovers inched their way inexorably up Pearl Pass in the reverse direction, a much trickier proposition than the trip two days before, then through the very narrow confines of Italian Pass. Friday's route included an old stagecoach route through Old Tincup Pass, but for the first two groups it was simply too wet to get through. After some fairly major rebuilding work on one particular 3ft rock step the last group of 10 managed it, and all three groups finished off their trek through Aspen's Independence Pass to the old mining town of St Elmo for a graduation dinner.

The full week had given most people their first taste of real off-roading, some 300 miles of it, and some said they would cheerfully have paid double for the experience. It had also demonstrated in no uncertain terms just what a well-driven Range Rover is capable of, and, despite the size of some of the rocks, damage was limited to a few lost bumper end caps and a couple of dented lower panels.

This August there will be another four courses, but with the level of over-subscription experienced in 1990 Tom Collins knows he will have to find some way of catering for more people: 'We are going to have to expand the courses, but frankly it's difficult to find a place in the States that has the draw of Aspen. We have looked at courses in Europe with Eastnor Castle as a second step, but having a real adventure as the final stage of the Driving Academy would really make the program – perhaps we'll organise an expedition in Africa'. Land Rover may have started its own driving courses at Solihull now, but the USA seems to be leading the way.

Even the boulders seem bigger...

One route included an old stagecoach run through Old Tincup Pass

PURELY ACADEMIC

Driving up the rear of the Aspen Mountains

The vehicles were limited edition 'Great Divide' Range Rovers

Photographs by Julian Mackie

HIGH ACHIEVERS

On road and off, Toyota's new Landcruiser and Mercedes' latest G-wagen tackle the Range Rover success story

When the Range Rover first saw the light of day back in 1970, it cost less than £2000. But the Range Rover then and the £34,450 Range Rover Vogue SE now are rather different animals, moving steadily upmarket over the years. And for most of that time no one else really bothered to break into the very specialised market niche Land Rover had managed to create. Most other four-wheel drive utilities were happy being neglected in muddy farmyards; none achieved that peculiar status which made the Range Rover as acceptable in Sloane Square as the Sahara.

For a long time the Japanese concentrated on creating a sub-Range Rover class. These less expensive 4wd estates had reasonable handling on and off road plus most of th creature comforts we've come to e pect from less adventurous saloon Land Rover's answer to this was th Discovery, and for a few months seemed that Range Rover's positic would remain unchallenged.

The Mercedes-Benz G-wagen ha the makings of being a Range Rov rival, but its original grimly militarist looks and 'soldier-proof' interior were bit off-putting. But while the ne G-wagen isn't exactly a 'chrysalis butterfly' act, the softer, less angul exterior and much improved interi have certainly changed its appeal.

Unlike the Range Rover, there a short- and long-wheelbase G-wagen with either 3.0-litre petrol or non-turl diesel engines which now permanent drive all four wheels. The long-whee base petrol 300 GEL costs £34,87 which includes anti-lock brakes – b

by the time all the extras fitted to the test model have been included, you are driving a £43,000 car . . .

If the last Toyota Landcruiser seemed at its happiest kicking up the dust along some African bush road, the latest version, the long-wheelbase VX, is very much more a creature of the Home Counties. While it has lost few of its off-road abilities, its distinctly 'mid-Atlantic' looks, with lots of curves and brightwork, have drastically changed its appearance.

Inside, too, everything is new, but the Landcruiser stays with diesel-only power – a full 4.2 litres of it – and, like the Range Rover and G-wagen, live axles and coil springs. The £28,646 price also reflects a move closer to Range Rover territory.

To confuse things a little, the much less expensive (£18,611) short-wheelbase Landcruiser II is a separate de-

MERCEDES-BENZ 300 GEL £34,870

Externally G-wagen now has softer lines but retains military bearing – until you step into wood and leather-laden interior. Dash's array of diff locks is for the rough stuff

RANGE ROVER VOGUE SE £34,450

Over 20 years old yet Range Rover is still king of off-roaders. Rolls less now, and performance is even better. Rather fiddly facia and narrow interior showing its age though

TOYOTA LANDCRUISER VX £28,646

Long-wheelbase Landcruiser looks brutish but in fact is easy to drive. Very car-like inside, extremely well equipped. Falls behind in off-road abilities, but most economical of trio

sign and not simply a downsized VX.

The Range Rover's familiar profile stays essentially unchanged, with manual transmission now available as a no-cost option to the standard four-speed automatic. With the increase in engine size from 3.5 to 3.9 litres, the V8 engine now packs a 185bhp punch, backed by a massive 235lb ft of torque.

If the handling was good previously, it's now been made better with the addition of a new anti-roll bar. Detail improvements mean a bigger fuel tank with the filler now mounted higher up on the rear wing, and a glass instead of steel panel for the sunroof. Soundproofing has also been improved and, with the addition of heated door locks, you shouldn't get frozen out of your Range Rover.

MERCEDES 300 GEL

Even wearing its new clothes, the G-wagen still retains its rather unforgiving, slab-sided military bearing, right down to the field green finish. It retains the air of a no-nonsense working vehicle, with bumpers you could push a tree down with. But open the door and you step up into a very different world – zebrano wood adorns the centre console and instrumer panel, the latter you might last hav seen in a Mercedes-Benz S-Clas saloon. And Mercedes is finally in ste with the rest of Europe: the singl indicator/wiper steering-column leve has been moved to the left.

Turn the key and the six-cylinde petrol engine sounds almost as swee as in any Mercedes saloon. And forge any idea that automatic transmissio and cross-country abilities don equate. The 300 GEL's four-spee automatic has a superslick change ar can be held in individual gears essential if you're heading right c road. There is a separate lever for th low-ratio transfer box which, with th front, centre and rear diff locks, give the 300 GEL huge versatility.

On the road, the Mercedes has well-balanced feel and the perforn

Luxury abounds in top Range Rover but is a five-seater only

ance is good, the 0-60mph increme coming up in 12.0sec. The brick-w aerodynamics limit top speed 102mph but it pays to keep speed dov on long journeys to help fuel consum tion. With our overall 14.3mpg, v needed the 20-gallon fuel tank. Th steering is very good for an off-road and, when you consider the relati simplicity of the suspension, both ri and handling are impressive.

On the rough stuff, the 300 GE handles everything you can throw at On slippery gravel and clay tracks th Mercedes' grip is good, while precip tous slopes can be tackled with cor plete confidence.

Drawbacks? The slightly dour loo apart, cost is the G-wagen's bigge problem. The basic price of £34,87 does include anti-lock brakes, but the time the test car's automatic tran mission, air conditioning, wide-rim a loy wheels and special wheel arche leather seats, radio/cassette and me allic paint have been added, the b tops a whopping £43,000. For th money, you'd have thought that th rear door might have been hinged the right instead of the left . . .

R/ROVER VOGUE SE

Turn the ignition key and you hear th burbling of the venerable but still wo derful V8 engine. The capacity is from the original 3.5 to 3.9 litres, a

HOW THE CARS COMPARE

	MERCEDES-BENZ 300 GEL	RANGE ROVER VOGUE SE	TOYOTA LANDCRUISER VX
PRICE	**£34,870**	**£34,450**	**£28,646**
PERFORMANCE			
Max in top (mph)	**102**	**108**	**102**
in 4th	—	—	96
in 3rd/2nd/1st	82/54/32	108/73/43	66/42/25
30-70 through gears (sec)	**12.4***	**11.5***	**14.0**
0-30	4.0	3.9	3.8
0-40	6.0	5.8	5.8
0-50	8.5	8.2	9.0
0-60	**12.0**	**11.3**	**12.3**
0-70	16.4	15.4	17.8
0-80	22.8	21.3	24.0
30-50 in gears 3rd/4th/5th	5.9	5.2	5.0/8.4/11.3
40-60	6.4	5.7	8.2/12.9/16.7
50-70	—/7.9/—	—/10.7/—	—/17.6/22.8
60-80	—/10.8/—	—/15.3/—	—/23.7/29.6
Top gear rpm/speedo at 70mph	3500/75	2600/74	2630/76
*In kickdown – auto transmission			
RUNNING COSTS			
What Car? test mpg	14.3	14.6	20.4
Touring mpg[1]	16.2	19.2	23.5
Gov't mpg: Urban/56/75	14.9/19.5/14.4	15.6/26.9/21.0	23.3/29,1/18.3
Fuel grade/capacity (galls)	UL/20.9	UL/18.0	diesel/20.8
Range at touring mpg	339	346	489
Insurance Group	**7**	**7**	**7**
Warranty (months/miles)	12/UL	12/UL	36/UL
Anti-rust (years)	—	—	6
What Car? cost per mile[2]	n/a	74.71	61.22
EQUIPMENT OPTION COST			
Power steering	yes	yes	yes
Central locking	yes	yes	yes
Electric windows	yes	yes	yes
Adjustable mirrors	elec	elec	elec
Anti-lock brakes	yes	yes	no
Sound system	no	yes	s.rad/cass
Alloy wheels	yes	yes	no
Split/fold rear seats	yes/yes	no/yes	yes
Leather upholstery	£2285	yes	no
Sunroof	£997	yes	yes
Adj steering column	no	no	yes
Adj lumbar support	no	no	yes
Auto transmission	£1370	yes	£1170
Air conditioning	£2578	yes	no
DIMENSIONS			
Length/wheelbase (in)	173/112	175/100	188/112
Width (inc mirrors) (in)	78	82	84
Height (in)	67	71	74
Headroom front/rear (in)	41/40	38/41	39/39
Legroom front/rear (in)	39-41/25-33	33-40/25-33	38-40/26-33
Rear shoulder room (in)	53	55	55
Boot capacity (cu ft)	47/91	36/71	28/—
Turning circle (ft)/lock turns	40.0/3.8	41.0/3.5	40.0/3.8
Kerb/towing weight (kg)	2220/2950	1988/4000	2405/3500
MECHANICAL SPECIFICATION			
Cyls/cc/fuel system	6/2960/inj	8/3947/inj	6/4164/diesel inj
Bore/stroke (mm)	89/82	94/71	94/100
Valvegear	sohc	ohv	sohc
Compression ratio	9.2:1	9.35:1	18.6:1
Power (bhp/rpm)	**170/5500**	**185/4750**	**165/3600**
Torque (lb ft/rpm)	173/4500	235/2600	266/1800
Brakes F/R	disc/drum	Vdisc/disc	disc/disc
Suspension front	live axle/coil	live axle/coil	live axle/coil
rear	live axle/coil	live axle/coil	live axle/coil
Tyres	205R16	205R16	265/75R15

[1]Calculated at 'Euromix' mpg (1/2 urban + 1/4 56mph + 1/4 75mph)

[2]Cost-per-mile figure calculated over three years and 36,000 miles. Includes fuel, depreciation, servicing and funding costs (figures supplied by Fleet Management Services Ltd, Shrewsbury)

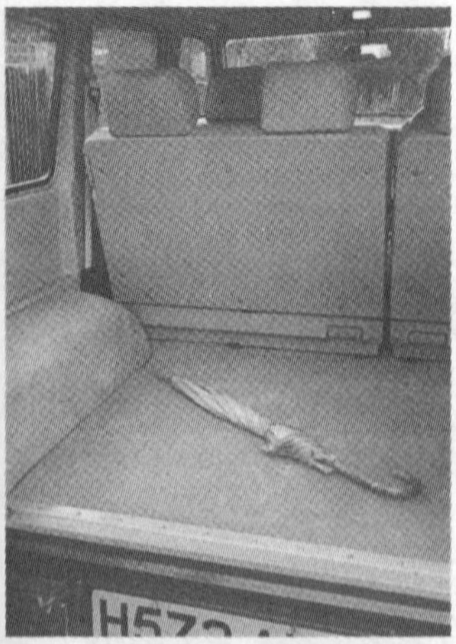

Wheel arches limit GEL's load area, door opens wrong way

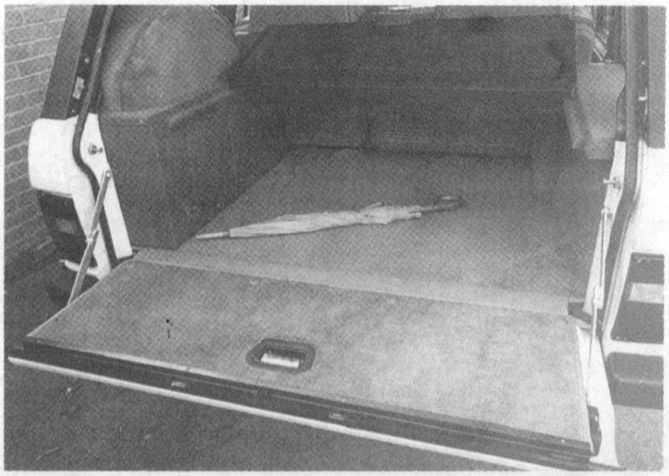

Range Rover retains split tailgate, spare takes valuable space

Roomy, comfortable Toyota's brown velour not to all tastes

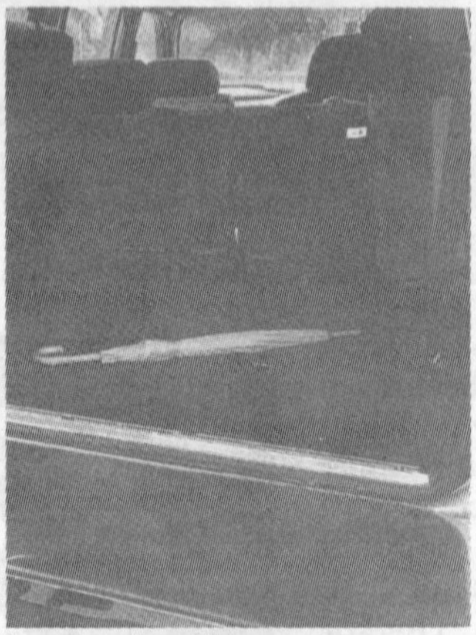

Landcruiser the most practical – a full seven-seater

now has fuel injection. Don't bother with the no-cost option manual gearbox: the change is heavy, the feel ponderous. The four-speed automatic is far more in keeping with the Vogue SE's top-of-the-market image.

There's no criticism of the driving position which, in estate agents' jargon, gives commanding views of the countryside. But the interior is starting to show its age, not helped by the cheap, downmarket quality of the plastic mouldings on the facia. The column levers have the same fiddly switches for the lights and wiper delay as a Metro, while most of the other buttons are tucked away under the main facia and hidden behind the steering wheel rim. In addition the seat-shaped switches for the electric front seat adjustment really do look as if they have been added as an afterthought.

On road or off road the Range Rover is still king. Despite its size and slightly woolly steering, the on-road handling is great, improved still further by the uprated suspension which helps reduce body roll in corners. While you don't treat it like a Mini, it can still be driven quickly and neatly.

Performance is good for a vehicle this size. It will storm to 60mph in 11.3sec and reach a top speed of 108mph. On the downside is the fuel consumption, which in everyday use will probably fall short of 15mpg. Rover has softened the blow a little by giving the Vogue an 18-gallon fuel tank.

Venture off road and that huge suspension travel means that traction is always there. A major plus point is that the ride too remains good, so that plunging in and out of steep-sided, water-filled 'bomb holes' can be performed at speeds which the 300 GEL or Landcruiser simply cannot match. The low-ratio transfer box provides that tremendous mud-plugging, hill climbing ability which is what the Range Rover is all about.

Room in the back is good but the Range Rover has stayed faithful to its original split tailgate, where the rear window section has to be lifted before the bottom section can be dropped flat. Awkward perhaps, but it does give an extended load platform which the other two don't have. The spare wheel, however, on the left-hand side of the load area, does take up a lot of space – and can be difficult to lift in and out.

TOYOTA LANDCRUISER

Deception is part of the Toyota's appeal. When you first see it, let alone climb into it, it has the appearance of a big, lumbering machine. This is not helped by the big flared wheel arches, running boards, sparkling chromium brightwork and the faired-in headlamps, hardly the stuff of muscle-flexing cross-country vehicles.

But don't let the Landcruiser's looks deceive you, because on the road it is a real mile-eater. The big 4.2-litre turbodiesel starts with a puff of smoke but the distinctive cold-start knock and rattle is well muted and soon fades. An automatic version is available, but the five-speed manual box is so slick and smooth it could have come from a Corolla. Yet the engine's massive torque – 266lb ft at just 1800rpm – means that you simply don't need to be constantly changing gear.

This big brute's no mean performer either, reaching 60mph from rest in just 12.3sec. Like the other two the top speed, at 102mph, is limited by the unaerodynamic shape. It is at its best on motorways, the engine hardly audible at 70mph, while in the lanes the sheer bulk of the VX tends to be the principal speed limiter. You are surprised to find country lanes you know well have become narrower overnight. This is not helped by the rather vague power steering, which makes precise placing difficult.

Off the tarmac you find the VX's limitations. The live-axle suspension, which was so good on hard surfaces, lacks the wheel travel needed in really demanding conditions, while the chubby Dunlop tyres may look good but don't have the cut-through-the-mud ability of the Range Rover's slimmer-section Michelins.

Try too hard and the VX has a habit of developing a vertical bounce, which means lost traction. The full-time 4wd system has front, centre and rear diff locks, with the centre one coming in automatically when low-ratio transfer gear is selected. But to be fair, just how many owners will ever take a vehicle like this that far off the beaten track?

Where the Landcruiser does score over its two rivals here is in fuel consumption. Despite its bluff-fronted styling, it still managed a very worthwhile 20.4mpg. With less time spent on motorways, the big Toyota may return close on 25mpg.

Inside, the VX is sheer civilisation. The facia is very car-like, with a sweeping cowl over the instruments and switches. Both front seats are heated, the driver's even having electric lumbar control. Separate heaters are provided for front and rear, plus electric windows and mirrors and headlamp washers.

Like the Mercedes 300 GEL and Range Rover, all-round visibility is great. Rear seat passengers get plenty of room and, when needed, two more people can be carried on the fold-away seats in the load area.

RATINGS AND VERDICT

	R/ROVER	MERCEDES	TOYOTA
PERFORMANCE AND ECONOMY	●●●○○	●●●○○	●●●○○
HANDLING AND RIDE	●●●●○	●●●○○	●●●○○
BEHIND THE WHEEL	●●○○○	●●●●○	●●●○○
ACCOMMODATION	●●●○○	●●●○○	●●●●○
QUALITY & EQUIPMENT	●●●○○	●●●●●	●●●●○
SERVICE AND COSTS	●●○○○	●○○○○	●●●○○
VERDICT	●●●●○	●●●○○	●●●○○

All these vehicles have undoubted off-road abilities but, at the price, would you take the risk? The Range Rover started life as a tough workhorse, but over the years its image has changed. Like the G-wagen, the ability is there, but there are less expensive vehicles which will tackle the rough and tumble every bit as well.

On road the Range Rover is the winner, with the G-wagen and perhaps surprisingly the Landcruiser not that far behind. But the Toyota does not feel at all happy as the hedgerows close in.

Venture off road and the Range Rover simply is the best – and not just of this trio. The G-wagen is a sturdy performer too, while the Landcruiser brings up the rear.

The Toyota will be reliable as Toyotas always seem to be, though it does fall down on its 'velour and brown plastic' image inside. The G-wagen has that feeling of being as near-perfect as a vehicle of this price and type needs to be. Everything works with machined precision: pity about the wrong-way-opening rear door, though. Alas, the Range Rover is starting to show its age, with more and more switchgear having to be found homes in and around the facia. The brilliant concept needs a major rethink.

At this price image is everything – and the Range Rover still has it. Instantly recognisable, with unstoppable off-road abilities, it simply remains the best. The G-wagen falls down on price and dour looks, while the Toyota has attempted a move up into company which it cannot quite match.

RANGE ROVER COUNTY SE

A SPECIAL EDITION FOR PEOPLE WHO DON'T LIKE OPTION BOXES

by Rik Paul

PHOTOGRAPHY BY CHARLIE RATHBUN AND SIDELL TILGHMAN

ROAD TEST Only a half dozen years ago, a four-wheel-drive sport/utility vehicle outfitted like a European luxury sedan and marketed for $35,000 to $45,000 seemed to have about as much a chance of succeeding as a Trabant dealer in Bel Air.

After all, according to prevailing conventional wisdom, the typical gravel-seasoned off-road enthusiast wouldn't touch a vehicle dressed in leather and burled wood, let alone pay more than the nation's median annual income for it. Similarly, to a typical, investment-minded luxury-car buyer, the thought of dust on his mirror-like finish is disturbing enough, never mind traversing rocks and streams.

Yet, over the last four years, Range Rover has tapped successfully into a buyer market that does indeed want the versatility of a serious all-terrain vehicle, without giving up the abundant creature comforts of a high-end luxury sedan. In fact, Range Rover has been successful enough to create a minor bandwagon for others to jump aboard, including the likes of Oldsmobile's new Bravada, the recent but ill-fated Laforza, and even the latest Toyota Land Cruiser incarnation.

After spending several days in a Rover following the Continental Divide through the Colorado Rockies a couple years ago, we can attest to the fact that the Range Rover has truly outstanding off-road capabilities. Ironically, though, the vast majority of Range Rover owners wouldn't be caught dead on anything more rugged than the freshly plowed road leading up to their favorite ski resort.

But that's because the Rover's attraction goes well beyond its all-terrain capabilities. It has panache. A high, see-over-traffic stance. A solid, German-luxury-sedan-like construction. A corrosion-resistant aluminum body. An excellent customer service program. Up-to-date engineering, including an innovative four-wheel anti-lock braking system. And of course, a pampering interior.

The Rover's 3.9-liter multipoint-injected aluminum alloy V-8 puts out 178 horsepower at 4750 rpm and 220 foot-pounds of torque at 3250 rpm. The engine is somewhat sluggish from a dead stop, due at least in part to having to motivate a full-time all-wheel-drive system through an automatic transmission. But once it overcomes inertia, the engine is smooth, quiet, and more than strong enough for easy passing, climbing hills, or high-speed cruising without a hint of strain.

The V-8 is mated to a four-speed automatic transmission, which provides cushioned, well-spaced shifts. From here, power is fed into a full-time four-wheel-drive system that includes a Borg-Warner viscous-coupled two-speed transfer case. While the Rover has always delivered a plush ride, thanks to its four-wheel

Far left: The County SE's custom-designed a/d/s/ audio system includes a subwoofer level control next to the Pioneer radio. Left: The CD changer, subwoofer enclosure, and amplifier are easily accessible without compromising valuable rear storage space.

coil-spring suspension, its on-road manners have been compromised by excessive roll in corners. For '91, the challenge was to reduce this without sacrificing the vehicle's excellent ride quality.

This was achieved by adding front and rear anti-roll bars, fitting softer bushings into the front trailing arms, dialing in softer damping characteristics into the shocks, and replacing the previous dual-rate rear coil springs with softer single-rate springs. The result is a 33-percent reduction in body roll without a noticeable compromise in ride quality. Which in real terms means that the obvious and annoying sway of previous models has been reined in, providing more stable, predictable cornering. One weakness still remaining is a somewhat dead on-center steering feel that leads to a touch of wandering during straight-ahead driving.

The Rover offers an impressive array of standard amenities, including eight-way power adjustable seats covered with Connolly leather, polished burled walnut trim, and heated door locks, windshield, and washer jets. Also standard are power headlight washers, cruise control, air conditioning, a two-stage auto-dimming rearview mirror, and power everything.

The County SE edition, at $47,300, retails for $4300 more than the base Range Rover and includes even more standard features: tilt-and-slide electrically operated sunroof, five-spoke alloy wheels, and a heavy-duty aluminum brush guard with removable headlamp guards. Although Range Rover offers no factory options for the SE, the vehicle we tested was also equipped with taillight guards, a $178 dealer accessory.

Perhaps there *should* be a few options, though. Such as a tilt steering wheel, map lights, drink holders, a full instrument cluster, and more storage space around the front seats. These are minor nits, for sure, but they're extras that are found in vehicles that cost far less.

The majority of the SE's price difference, however, is due to an a/d/s/ audio system specially designed for Range Rover to rival the quality of high-end home stereo systems. The

heart of the system is a Pioneer in-dash AM/FM/cassette player that includes a weatherband channel, and also controls a six-disc CD changer mounted in the rear compartment. A 150-watt six-channel a/d/s/ power amplifier provides more than enough punch. And nine premium speakers, including an innovative subwoofer enclosure, make it one of the best factory-installed sound systems we've heard.

In the final assessment, the County SE delivers. It gives you a strong, smooth, reliable powerplant, excellent ride quality, improved handling, an advanced drivetrain, and impeccable ergonomics. And for those few who really care, it'll go just about anywhere you want. **MT**

SECOND OPINION

Somehow, it's difficult to picture a couple in formal attire brush-busting their way up the Mojave Trail. Cruising to the overpriced grocery clinic is more in the County SE's league. It's a shame, too, because with its full-time four-wheel-drive system, four-wheel coil-spring suspension, and sturdy V-8 engine, it's an admirable off-road runner.

Inside, the SE could double for a limousine. The sound system reaches even beyond *my* need to make my eardrums bleed, and the seats welcome you to settle in for the long haul. Sure, some things are missing, but, hey, if you can afford to spend close to $50 grand on a vehicle like this, you'll already have those practical items inside your other 12 cars.

—B.J. Hoffman

TECH DATA

Range Rover County SE

GENERAL

Make and model..........Range Rover County SE
Importer............Range Rover of North America, Inc., Lanham, Md.
Body style..........................4-door, 5-passenger
Drivetrain layout..............................Front engine, four-wheel drive
Base price..$47,300
Price as tested..$48,053
Options included............Taillamp guards, $178; destination, $575
Typical market competition............Toyota Land Cruiser, GMC Suburban, Neiman-Marcus account, Bertram yacht, condo in Aruba

DIMENSIONS

Wheelbase, in./mm..........................100.0/2540
Track, f/r, in./mm..............................58.5/58.5/ 1486/1486
Length, in./mm...................................175.0/4447
Width, in./mm......................................71.4/1813
Height, in./mm.....................................70.8/1798
Ground clearance, in./mm........................8.0/204
Manufacturer's curb weight, lb...................4401
Weight distribution, f/r, %........................49/51
Cargo capacity, cu ft....................................36.2
Fuel capacity, gal..21.6
Weight/power ratio, lb/hp...............................24.7

ENGINE

Type...V-8, liquid cooled, cast aluminum block and heads
Bore x stroke, in./mm.......................3.70 x 2.80/ 94.0 x 71.1
Displacement, ci/cc...........................241/3947
Compression ratio...8.1:1
Valve gear......................OHV, 2 valves/cylinder
Fuel/induction system..................Multipoint EFI
Horsepower,
hp @ rpm, SAE net......................178 @ 4750
Torque,
lb-ft @ rpm, SAE net....................220 @ 3250
Horsepower/liter...45.1
Redline, rpm..6000
Recommended fuel..............Unleaded premium

DRIVELINE

Transmission type..........................4-speed auto.
Gear ratios
(1st)..............2.48:1
(2nd)..............1.48:1
(3rd)..............1.00:1
(4th)..............0.73:1
Axle ratio...3.54:1
Final drive ratio..3.11:1
Engine rpm,
60 mph in top gear.....................................2200

CHASSIS

Suspension
Front...............................Live axle, leading arms, Panhard rod, coil springs, anti-roll bar
Rear........................Live axle, lower links, upper A-arm, coil springs, self leveling, anti-roll bar
Steering
Type....................Worm and roller, power assist
Ratio..19.3:1
Turns, lock to lock..3.4
Turning circle, ft..39.4
Brakes
Front, type/dia., in................Vented discs/11.8
Rear, type/dia., in..............................Discs/11.4
Anti-lock..Standard
Wheels and tires
Wheel size, in...16 x 7.0
Wheel type/material....................Cast aluminum
Tire size...205/R16
Tire mfr. and model..........Michelin X M+S 244

INSTRUMENTATION

Instruments....130-mph speedo; 6000-rpm tach; fuel level; coolant temperature; analog clock
Warning lamps....................Oil; coolant; battery; ABS; washer fluid; check engine; rotate brake; brake temperature; tow

PERFORMANCE AND TEST DATA

Acceleration, sec
0-30 mph...3.8
0-40 mph...5.7
0-50 mph...8.2
0-60 mph...11.0
0-70 mph...15.3
0-80 mph...20.9
Standing quarter mile,
sec @ mph..................................18.2 @ 76.1
Braking, ft
30-0 mph..31
60-0 mph..136
Handling
Lateral acceleration, g..............................0.72
Speed through 600-ft
slalom, mph...56.9
Speedometer error, mph

Indicated	Actual
30	28
40	38
50	47
60	57

Interior noise, dBA
Idling in neutral..56
Steady 60 mph in top gear...............................70

FUEL ECONOMY

EPA, city/hwy., mpg.................................13/16
Est. range, city/hwy., miles............281/346

first **DRIVE**

Range Rover Vogue LSE

- *Flagship model with electronic traction control*
- *4.2-litre, V8 engine, 200bhp, 0-60 9.9sec*
- *Price £40,000. Motor Show launch*

SIMPLY THE BEST

● ***King off the road — new Range Rover is a case of the best getting better***

'You'll never get that up there. I've never seen the track as bad as this.' The driver of the battered, leaf-sprung Land Rover stared wide-eyed at £40,000-worth of brand new Range Rover Vogue LSE. He reckoned it had been tricky coming down the jagged rock steps, made slippery by the torrential rain — now we were planning to climb up them.

As our local advisor parked up to watch the fun, a press of a button lifted the Range Rover another 40mm on its air suspension. Sliding the transfer lever forward into low range and the gearshift back into D, we set off up the jagged rock steps. The deep burble of the new 4.2-litre V8 hardly changed pitch as the long wheelbase LSE walked effortlessly up, axles articulating to keep the wheels firmly on the ground and the Range Rover's underside well clear of the biggest boulders.

Two decades after it first saw the light of day, the Range Rover has undergone its most significant changes ever. A new flagship, the Vogue LSE, has an 8in-longer wheelbase, air suspension, a bigger V8 engine and electronic traction control. With leather, wood, electric everything and the best standard-fit stereo I've ever heard, we are talking serious luxury here — even with the front seats fully back, there is more legroom in the rear than in the long-wheelbase versions of both the Mercedes S-Class and the BMW 7-Series — precisely the cars that the Vogue LSE is meant to compete with.

Of course, most Range Rovers never get near a Welsh mountain, so it's important to point out that, with 200bhp and 250lb ft of torque on tap in its new 4.2-litre, 3-way catalyst guise, the long-serving Rover V8 has huge helpings of mid-range. It's also more impressive than ever on the road. Despite the weight increase — at 4739lb the LSE weighs 308lb more than a standard Vogue SE — the new engine slices half a second off the 0-60 time, down to 9.9 seconds in four-speed automatic guise (there's no manual option) and gearshifts are as seamless as ever on part or full throttle. Top speed is still limited to 112mph to keep within the tyres' speed rating, but the LSE is perfectly happy cruising at 100mph, the V8 just a muted rumble.

Easily the biggest leap forward comes from the electronically-controlled air suspension (EAS), which uses air springs and a sophisticated computer control system to give five different ride heights, as well as automatic self-levelling both front to rear and side to side. The standard setting mirrors the height of the original coil-spring suspension, but the system automatically switches to low profile, lowering the Range Rover by 20mm, when cruising above 50mph. For off-roading a high profile setting gives an extra 40mm of ground clearance, and if the system senses that the bottom of the Range Rover is still grounded out on an obstacle it can dial-in an extra 30mm of travel at each wheel to lift it clear again. Finally, the suspension can be lowered right down to 'access' mode, with the vehicle at a standstill.

The air springs deliver a significantly more supple ride than before, yet the LSE rolls less and demonstrates far better control through a series of S-bends than its coil-sprung predecessor. On dual purpose 205 R16 Michelin XM+S 244 tyres it also has plenty of grip, just nosing gently wide on the limit and delivering outstanding traction in the wet, thanks to its sophisticated, viscous coupling-controlled permanent 4wd system. Drive at more than 50mph for 30 seconds, and the EAS automatically lowers itself into low profile, improving the steering geometry and making the LSE feel tauter, more stable and with a degree of steering feel never possible in a Range Rover before.

As well as making light work of even really serious mud and rocks, the Range Rover floats across fields or along rough tracks, its off-road ride in a different league even to the world-beating performance of the coil-sprung version. The high profile setting gives even the long wheelbase vehicle (EAS is available on more humble Range Rovers too) a useful advantage in terms of ground clearance, and during a full day's off-roading in the Welsh mountains, it never needed to show off its extended mode — in fact it didn't touch even the front spoiler or towbar down once. A 12inch rock step did demonstrate the advantages of the electronic traction control — we clambered up after ETC had applied the brake to a spinning rear wheel. Earlier, we had also put it to the test climbing a steep, muddy river bank.

Faster, smoother, more refined and better handling than ever on the road, this Range Rover has lost none of its legendary ability on the rough stuff — in fact, the benefits to ground clearance, traction and ride comfort off-road are even more striking than the improvements on tarmac. In the case of the Range Rover LSE, the best just got better.

Howard Lees

- **FOR** Almost everything
- **AGAINST** Hardly anything
- **VERDICT** Simply the best in its class

● ***4.2-litre V8 engine produces 200bhp with a muted burble***

Range Rover Vogue LSE

PETER BURN

The ultimate Range Rover is a better off-roader than ever. Can it be a £40,000 luxury car, too?

IF WE WERE NOT ALL used to the Range Rover, its very notion would be greeted with great suspicion. Even in 1992, it would seem an impossible brief to produce a car that is not only perhaps the most capable off-roader in the world but also one so effective on the bitumen that the vast majority of its owners would never feel the need even to muddy its wheels.

But Land Rover's flagship has been evolving for 22 years now. And, as in any good democracy, it is that majority to whom the bulk of the legion changes found within this Range Rover LSE have been addressed.

The customers clamoured for more space inside, so they were given an extra eight inches, all within the wheelbase. This flagship could not be allowed to be slower than a lesser Range Rover, so the old V8 war-horse under the bonnet expanded its internal dimensions one more time to reach nearly 4.3 litres, up from the 3.9 litres of old. It had to be more comfortable too, if a still higher price tag was to be swallowed. So no longer does the Range Rover ride on steel springs, instead it floats on air suspension and allows the driver to vary the ride height from the cockpit.

The interests of the mud-plugging minority have not been ignored either. A traction control system has been installed to control wheelspin by activating the rear brakes when it detects slip.

The result is just what Land Rover wants; a car that's impossible to pigeonhole. Its £39,995 list price punches a hole in the off-roader stratosphere. Even the most expensive version of the only car ever to attempt to play the Range Rover game, Mercedes-Benz's 300GEL, looks cheap by comparison at £33,529. BMW's 525i Touring applies a different solution to the problem of being both utilitarian and luxurious at the same time, yet still comes up with an answer that's nearly £13,000 less than the Range Rover. No, if you want to find its real rivals you need to wheel out limousines like ▶

Price as tested £40,694 **Top speed** 110mph
0-60mph 10.8secs **MPG** 16.7

For Cabin space, smooth transmission, fabulous off-road, charm
Against Double-take price, wind noise, switchgear, low-speed ride

GE ROVER
K89 YKV

◀ the Lexus LS400 or Jaguar Sovereign, and even these contrive to cost several thousands less than Solihull's newest hero.

Can any car, designed in the '60s, with live axles at each end, be worth so much? Or has Land Rover finally asked too much of its noblest warrior and sent it over the top with an even heavier back-pack to fight younger, fitter opponents armed with a massively more sophisticated array of weapons?

Performance

As in so many other ways, the Range Rover's performance lands it plumb in the no-man's land between the class it has outgrown and that to which it still aspires. The LSE is not just the most powerful, fastest Range Rover of all. It outguns and outruns every other standard off-roader on sale in the UK too. Its 4278cc engine retains the old pushrods and two-valves per cylinder but you now have 200bhp at 4850rpm and 250lb ft of torque at 3250rpm, up from the 185bhp and 235lb ft wielded by the 3.9-litre engine which resides, unchanged, in the standard wheelbase Range Rover.

That said, you would have to resort to the test track to tell the difference between the two. The LSE weighs 2079kgs, some 69kgs more than its sister, meaning a scant power to weight ratio improvement of 4bhp per tonne to 96bhp per tonne. Not only does the LSE not feel fast, it doesn't sound fast either. The sweet V8 howl that has earned this engine such fame and fortune over the years has suffered at the hands of its long-throw crankshaft. Now it roars readily enough but becomes hoarse as its 5500rpm limiter approaches and merely undistinguished at more modest revolutions.

Not only can't it hold a candle-light to the performance of those rivals which spend all their time on tarmac, its acceleration is also shy of Land Rover's expectations. Far from being the first off-roader to reach 60mph in single figures as alleged by its manufacturer, the LSE actually takes 10.8secs, an identical time to the standard Vogue SE. Nor could it reach its claimed 112mph maximum, as full throttle simply rammed the engine into its limiter in third gear. Holding the pedal slightly off the floor resulted in an average 110mph in top. Once it has picked up its two tonne-plus girth, its performance improves dramatically, the 10.6secs it takes to reach 70mph from 30mph is more than half a second faster than the 3.9-litre version.

At least its transmission fits this bill, or indeed that of a more conventional luxury car. For a car driving all its wheels, the absence of any slack or shunt in the process is remarkable. Change-up points are well judged and slickly executed. You have to press hard to downchange and the response to such a request is by no means immediate, but at least the shift itself is smooth.

High compression pistons have been conscripted to improve fuel economy and have done an impressive job. Overall it returned 16.7mpg over 1000 miles, and approaching 20mpg over our touring route. One tankful that comprised mainly urban and country road driving saw consumption fall to nearly 14mpg but even this is no disaster for one so heavy, powerful and aerodynamically inefficient.

Handling

Even with air springs and anti-roll bars at each end, it's no good leaping into the LSE and charging down a favourite road expecting it to handle like a normal car. Its suspension is not independent, it is shod with dual purpose Michelin Mud and Snow tyres, and the springing, though a little firmer than the steel-sprung version, is still resolutely soft. The amount of body roll you can generate is still spectacular even though the anti-roll bars have controlled the rate of roll effectively. But once you have learned that the Range Rover's handling is merely different, not bad, you can hustle it along at a fine rate. The trick is to be as smooth as possible. The steering, a little jittery about the straight-ahead, smooths out when you apply lock and allows you to turn into corners as accurately as you could expect given the tall-walled 205/R16 tyres. So long as you don't wrench the helm over, the body rolls without lurching right up to surprisingly high limits. Should the front end start to slip, it will do so gradually and should allow you more than enough time to slow to a more sensible speed.

Off road it is roundly better than ever. Engine braking with the larger capacity, high compression engine is awesome, making the steepest, most slippery slopes look like scarcely a challenge. Going up the other side, the grip afforded by the traction control, all-wheel drive and phenomenal axle-articulation makes you think the wheels have transmogrified into caterpillars. Those who think the extra 8ins length must have compromised ground-clearance have not reckoned with the air-suspension, which allows you to raise the body fully 130mm from its lowest setting. It's a shame that so many owners will never discover what treasures await off the beaten track.

Ride

The suspension's new-found firmness is usually welcome, especially as it all but eliminates float at high speed. What it has not been able to do, though, is start to bridge the gorge between the kind of suspension refinement Lexus and Jaguar drivers take for granted and the rather more agricultural ride tolerated by those who drive off-roaders. Of this breed, the Range Rover is better than most but that does not stop it jolting over pot-holes in town and shuddering over transverse ridges.

Brakes

This is a perennial headache for those who drive large, all-terrain cars. They are heavy, yet need narrow tyres and both factors legislate against the ▶

Driving in the air — the LSE suspension

SUSPENDING A CAR AS HEAVY AND AS HIGH AS A Range Rover has never been easy. But by replacing the coil springs with air springs and rubber dampers, Land Rover reckons its new air suspension system (EAS) has solved the Range Rover's body control problems, and for the most part we'd agree. Adjustable to the tune of five different ride heights — standard, low profile, high profile, access mode (which drops the car 60mm below the ordinary ride height) and extended profile (which raises it 30mm above) — the Range Rover's air suspension acts similarly to a self levelling set up. However, it also lowers the body by 20mm above 50mph to provide better stability and steering response, and raises it by 30mm if the vehicle becomes grounded. Standard on the LSE but optional on all other petrol versions for around £1500, EAS is a must and should be standard across the range.

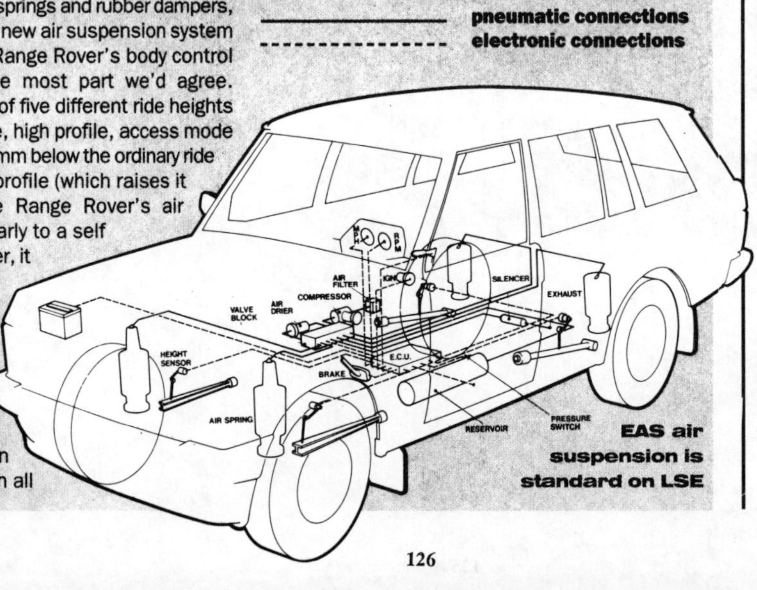

EAS air suspension is standard on LSE

Range Rover Vogue LSE

You can't help being impressed by the airiness and opulence of the Range Rover's cabin Electric front seats are exceptionally comfortable over long distances, even though the steeply raked wheel is unadjustable and too far away from the driver

The extra legroom in the back makes most alleged luxury cars look cramped in comparison. Four large adults can travel in space and, thanks to the self-levelling suspension, fine comfort. Rear central seat is too hard. Boot is small until you fold down the back seats

New 4.2-litre engine gutsy and not too thirsty but aurally uninspiring. Dash looks old, gives little information and is difficult to read at a glance. Switchgear muddled

Standard fit cruise control is just part of a good range of equipment which goes some, if not all the way to justifying the list price

After 22 years, the Range Rover's shape has aged but still looks great

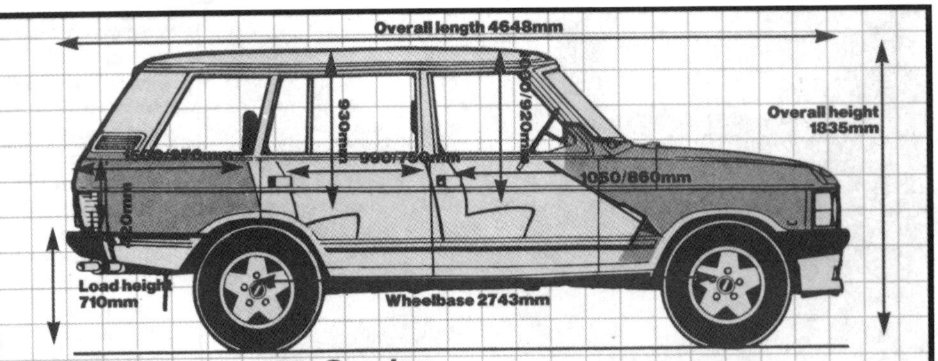

Width 1778mm
Front/rear tracks 1486/1486mm
Turning circle 12.7m
Interior width front/rear 1480/1500mm
Max/min boot width 1100/1050mm
VDA boot volume 1.02cu m
Cd 0.45

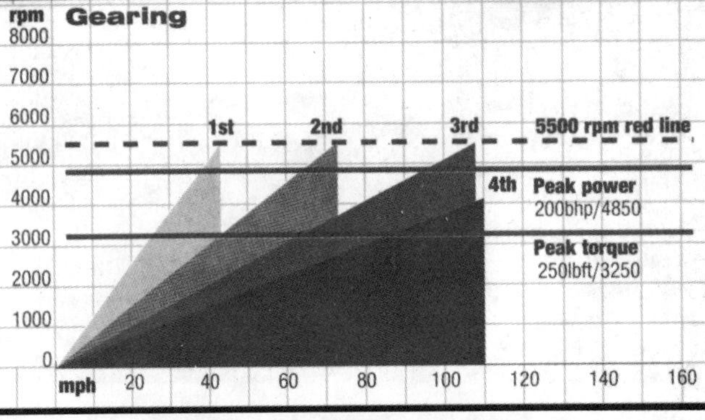

Specification

Engine

Layout 8 cyls in vee, 4278cc
Max power 200bhp at 4850rpm
Max torque 250lb ft at 3250rpm
Specific output 47bhp/litre
Power to weight 96bhp/tonne
Installation longitudinal, front, permanent four-wheel drive
Made of all aluminium
Bore/stroke 94/77mm
Comp ratio 8.9:1
Valves 2 per cyl, ohv
Ignition and fuel electronic ignition, Lucas L-Jetronic fuel injection, catalytic converter

Gearbox

Type 4-speed automatic
Ratios/mph per 1000rpm
1st 2.48/7.9 **2nd** 1.48/13.2 **3rd** 1.00/19.6 **4th** 0.73/26.9 **Final drive** 3.54:1 **High/low ratios** 1.21/3.24

Suspension

Front live axle, radius arms, air springs, anti-roll bar **Rear** live axle, radius arm, air springs, anti-roll bar

Body

Type 5dr, unstressed aluminium and steel panels, steel box section chassis

Steering

Type power assisted worm and roller
Lock to lock 3.8 turns

Brakes

Front 298mm ventilated discs **Rear** 290mm plain discs **Anti-lock** standard

Wheels and tyres

Size 7x16ins **Made of** alloy **Tyres** 205 R 16 Michelin XM+S **Spare** full-size

Made and sold by

Land Rover Ltd, Lode Lane, Solihull, West Midlands B92 8NW Tel: 021 722 2424

Performance

Maximum speeds

Top gear 110mph/4090rpm
3rd 108/5500 **2nd** 73/5500
1st 43/5500

Acceleration from rest

True mph	Secs	Speedo mph
30	3.8	32
40	5.6	42
50	8.0	52
60	10.8	62
70	14.4	73
80	19.7	83
90	26.3	94
100	38.4	104

Standing qtr mile 18.1secs/77mph
Standing km 33.1secs/96mph
30-70mph through gears 10.6secs

Acceleration in kickdown

mph	secs	k/down gear
10-30	2.9	1st
20-40	3.3	1st
30-50	4.4	1st/2nd
40-60	5.2	2nd
50-70	6.4	2nd
60-80	8.9	2nd/3rd
70-90	11.9	3rd
80-100	18.7	3rd

Fuel consumption

Overall mpg on test 16.7
Best/worst on test 19.2/9.6
Touring 19.2
Range 346 miles
Govt tests (mpg):
urban 13.1 **56mph** 24.2 **75mph** 18.7
Tank capacity 82 litres (18 galls)

The figures were taken at the Lotus proving ground, Millbrook, with the odometer reading 2500 miles. Autocar & Motor test results are protected by world copyright and may not be reproduced without the editor's written permission

Brakes

Distance travelled under max braking
Track surface dry
Anti-lock standard

30mph	11.3m
50mph	33.8m
70mph	66.2m
st qtr mile (77mph)	85.0m

Fade tests
Consecutive brake applications at 0.5g retardation from st qtr terminal speed

(figures on the right represent pedal pressures)

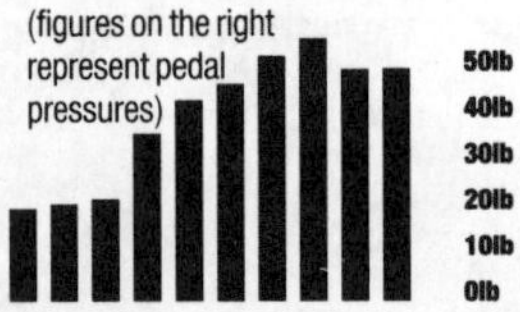

Weight

Kerb (incl half tank) 2079kg
Distribution f/r 50/50%
Max payload 541kg
Max towing weight 4000kg

What it costs

Prices

List price £39,995
Total as tested £40,694

Equipment

(prices in **bold type** denote option fitted to test car)

Automatic gearbox	★
Metallic paint	★
Seatbelt pre-tensioners	–
Variable speed intermittent wipe	★
Electrically adjustable mirrors	★
RDS radio/cassette	★
CD player	**£699**
6 speakers	★
Alloy wheels	★
Adjustable steering column	–
Electric windows all round	★
One-shot electric driver's window	★
Central locking on any door	★
Height/tilt adjustable driver's seat	★
Driver's seat lumbar adjustment	–
Electric seat adjustment	★
Rear central arm rest	★
Anti-lock brakes	★
Anti-theft system	★
Full leather trim	★
Traction control	★
Trip computer	–
Heated seats	★
Air conditioning	★

★ standard – not available

Insurance group n/a

Depreciation 50

Value after three years/30,000 miles as a percentage of new price. Source: Fleet News Guide to Future Residual Values

Warranty

12 months unlimited mileage, 12 months recovery. No anti-corrosion warranty

Servicing

Major 12,000 miles, service time 4.5hrs
Interim 6000 miles, service time 2.2hrs

Parts costs

Oil filter	£14.39
Set spark plugs	£8.27
Brake pads front	£48.76
Brake pads rear	£40.77
Exhaust (inc cat)	£271.39
Door mirror glass	£78.94
Tyre (each, typical)	£124.79
Windscreen	£397.16
Headlamp unit	£93.94
Front wing	£234.47
Rear bumper	£118.12

1 Cigarette lighter **2** Ventilation/air conditioning controls **3** Clock **4** Transfer box **5** Air suspension control **6** Fog lamps, interior light and cruise control switches **7** Electric mirror adjustment **8** Water temperature gauge **9** Fuel gauge **10** Light switch **11** Hazard lights **12** Cruise control set/reset **13** Variable intermittent wipe **14** Windscreen wash/wipe control

Range Rover can still roll prodigiously but cornering is much more secure than it looks. Even right on the limit...

◀ braking system. The LSE's large brakes and anti-lock help assure total stability under panic braking and provide acceptable resistance to fade, but what they don't do is slow the car quickly. A Lexus, for example, needs 10 metres less to stop from the legal limit and a BMW 525i Touring stops over 10 per cent more swiftly from the same speed.

The feel of the brake pedal is more a niggle than a concern. It is over-sensitive and has only very limited travel. Any Citroen XM driver will be familiar with this problem and, more than likely, tell you that it disappears as you adapt. However, it still requires concentration to bring the Range Rover to a smooth and measured halt.

At the wheel

★★★ It would be difficult to overstate the advantages of having a driving position as high as the Range Rover's. It goes beyond the imperious feeling gleaned from looking down on your fellow motorist. More importantly, it allows you to see over the top of them, and spot trouble vital seconds before drivers of more conventional cars. Indeed, all-round visibility is marvellous. Even so, the driving position will not be perfect until an adjustable steering wheel is fitted, as it is still too far away from the driver.

Like the Porsche 911 and others that have cheated the hang-man for longer than their makers ever imagined, its age shows on the inside. Different styles have been allowed to evolve and extra switchgear shoved where there's room. The ventilation controls are difficult to reach, tricky to understand and noisy to operate. A bank of switches to the right of the fan levers preside over all operations from activating the cruise control past the heated windscreen and fog lamps to ride height. They are out of your natural sight line and a confusing jumble to use. Even the instruments look like '70s refugees and are far from the last word in clarity.

Accommodation and comfort

★★★★★ If you want to see where the extra 200mm in the wheelbase has been spent, open the large rear door (it will swing through nigh-on 90deg) and take a look. Regardless of whether the front seats are all the way forwards or backwards, the Range Rover provides an extra 120mm more rear legroom than a Lexus with a similarly-set front seat. For those over 6ft, that's the difference between cramped and comfortable. Even with the driver's seat set all the way back, four such sized adults can travel hundreds ▶

...all four wheels stay on the ground, thanks to anti-roll bars

It may not appear 8ins longer but one look at the rear door shows where the extra length has been found

◀ of miles with stretching space to spare.

Headroom front and back is not quite so expansive but it impresses nonetheless. Where the Range Rover LSE scarcely benefits from its new-found length at all is in its boot, which remains disappointing short for so long a vehicle.

Despite the flawed driving position, the Range Rover remains a comfortable means of travel, thanks in the main to electric seats that prove brilliantly supportive over long distances even though they do allow you to slide around a little in the fastest corners.

Noise

★★

For a car tilting at the Lexus, the Range Rover is almost unforgivably noisy. Its noise levels are those of a family hatchback, not a luxury saloon. Unlike the hatch, though, the noise is mainly wind, not engine derived. Now there is air rather than steel between you and the axles, the tunnel of road noise from the tyres to the cabin has been successfully walled and long gearing keeps the engine quiet. But the LSE's bluff shape (its drag-coefficient is 0.44 even on the suspension's lowest, most aerodynamic setting) means you never escape from the sound of wind on the motorways.

Build quality and safety

★★★

You do not expect to find such wide panel gaps, such cheap plastic and wood veneer on so expensive a car. The paintwork on the test vehicle had also dulled after only a few months' work.

Yet, despite all these visible faults, the LSE feels solid and strong enough to protect you in an accident that would destroy many a lesser vehicle. Certainly the only creak we could detect over our 1000 mile test came from the comforting sound of the leather upholstery.

Value and equipment

★★★

Acres of leather, air-conditioning, cruise control, heated seats, a burglar alarm and traction control are just some of the highlights of a comprehensive standard equipment list that lacks only a trip computer and an adjustable steering column. Even so, it is still difficult to say the name Range Rover and £39,995 in the same breath.

It is almost as if Land-Rover is dipping its toe into the market water to see just how hot it is; after all, it doesn't replace any other Range Rover models, merely tops the range. Only time will tell if the LSE can take the heat.

Verdict

★★★

The Range Rover continues to confound convention. Convention says luxury cars and off-roaders should represent contrasting market poles so utterly different are their aims. That the Range Rover manages to play either part with any conviction is impressive enough. With its bigger engine, bigger cabin and swanky suspension it seems set to maintain, even enhance, its appeal to those who, in their droves, buy a Range Rover not for the superlative off-roader that it is, but for the luxury saloon it tries heroically but ultimately fails to be.

Any assessment of performance, ride or refinement reveals that the likes of the Lexus and Jaguar are simply light years ahead of the Range Rover. Yet, like that king of automotive Peter Pans, the Porsche 911, its seniority has earned it the right to convert its faults into charm, letting the usually infuriating be perceived as the quirkily endearing. And there has never been a Range Rover that has been either as simple to fault or as easy to like as the LSE. So its success may not be deserved, but we, for one, will not begrudge it.

Range Rover LSE

★★★ £39,995

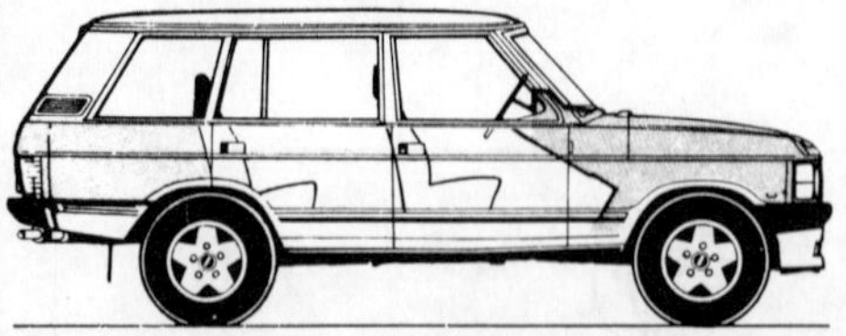

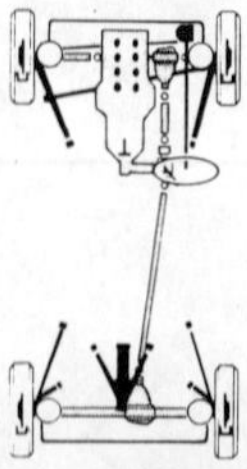

Length 4648mm **Width** 1778mm **Front track** 1486mm **Wheelbase** 2743mm
Height 1835mm **Rear track** 1486mm **Weight distribution front/rear** 50/50

A longer body, more powerful engine and air suspension pushes this Range Rover's price into uncharted territory for off-roaders. In mud it is better than ever but on the road mild performance, an unconvincing ride and poor refinement show its age like never before. That it manages to remain so likeable despite these faults is a fine testament to a great car.

Capacity	4278cc
Power	200bhp/4850rpm
Torque	250lb ft/3250rpm
Max speed	110mph
0-60mph	10.8secs
30-70mph through gears	10.6secs
Standing quarter mile	18.1secs
30-50mph in kickdown	4.4secs
50-70mph in kickdown	6.4secs
Mpg overall/touring	16.7/19.2
Mph/1000rpm in top	26.9
Kerb weight	2079kg
Date tested	14.10.92

BMW 525i SE Touring

★★★★ £27,350

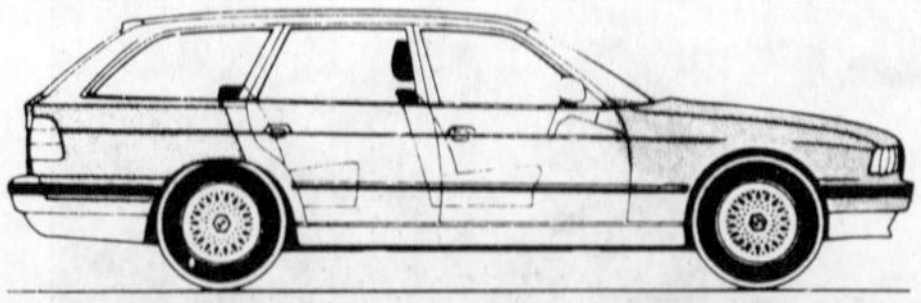

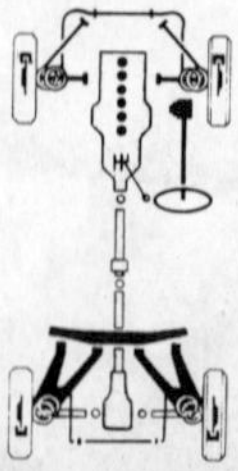

Length 4535mm **Width** 1690mm **Front track** 1465mm **Wheelbase** 2475mm
Height 1290mm **Rear track** 1465mm **Weight distribution front/rear** 53/47

It follows that one of the most classy range of saloon cars available should translate into an an estate that's equally desirable and largely unhindered by its modest bulk. The Touring gives a little performance away to the saloon but still handles just as well. Neatly presented and well packaged, its load area should prove fine for most people this side of needing Volvo-sized boots.

Capacity	2494cc
Power	192bhp/5900rpm
Torque	180lb ft/4700rpm
Max speed	132mph
0-60mph	9.2secs
30-70mph through gears	9.2secs
Standing quarter mile	17.0secs
30-50mph in kickdown	9.8secs
50-70mph in kickdown	14.2secs
Mpg overall/touring	20.8/25.8
Mph/1000rpm in top	22.9
Kerb weight	1590kg
Date tested	1.4.92

Jaguar 4.0 Sovereign

★★★★ £38,300

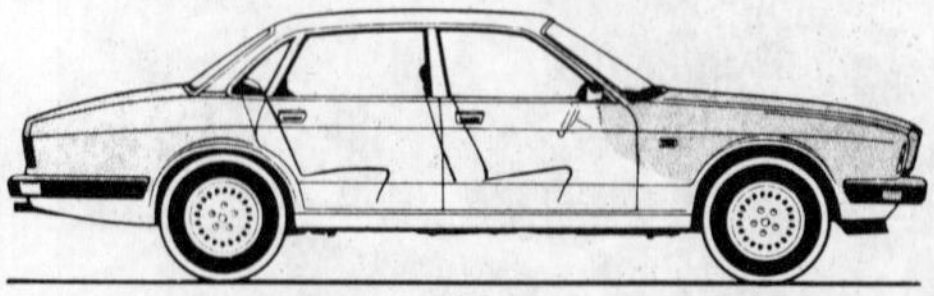

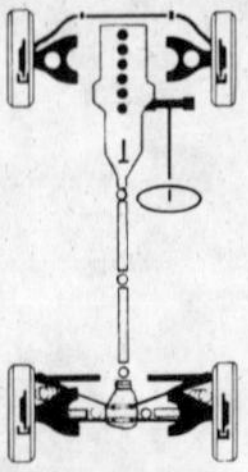

Length 4988mm **Width** 2015mm **Front track** 1500mm **Wheelbase** 2870mm
Height 1380mm **Rear track** 1498mm **Weight distribution front/rear** 54/46

OUR CHOICE

Now on sale, the 1993 XJ6 may look unchanged but beneath the skin a minor revolution has taken place. A recalibrated gearbox and new camshafts means progress is even quieter, an airbag is standard and many ergonomic improvements make the car easier to operate. For its suspension control and unaffected class, it's still a marvel.

Capacity	3980cc
Power	223bhp/4750rpm
Torque	278lb ft/3650rpm
Max speed*	139mph
0-60mph*	8.3secs
30-70mph through gears*	7.3secs
Standing quarter mile*	27.8secs
30-50mph in kickdown*	n/a
50-70mph in kickdown*	4.4secs
Mpg overall/touring*	18.5/23.8
Mph/1000rpm in top*	29.2
Kerb weight*	1750kg
Date tested	13.9.89

* Figures for non-catalysed car

Lexus LS400

★★★★ £36,991

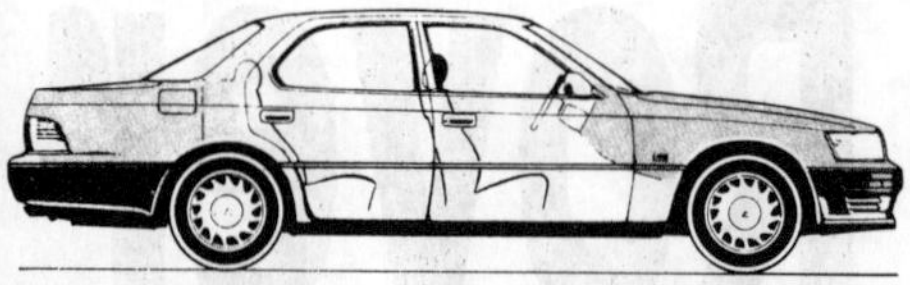

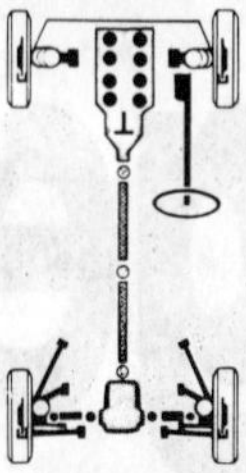

Length 4995mm **Width** 1820mm **Front track** 1565mm **Wheelbase** 2815mm
Height 1425mm **Rear track** 1555mm **Weight distribution front/rear** 54/46

It has the quietest engine, best low-speed ride and best overall refinement of any car we have driven. It is a technical *tour de force*. But criticisms can be levelled at its undistinguished looks and apparent inability to make its owner feel special — something that a Jaguar Sovereign excels at. Outhandled by the new BMW 740i, the Lexus is still fabulous value for money.

Capacity	3969cc
Power	241bhp/5400rpm
Torque	258lb ft/4400rpm
Max speed	136mph
0-60mph	8.6secs
30-70mph through gears	7.1secs
Standing quarter mile	16.3secs
30-50mph in kickdown	3.0 secs
50-70mph in kickdown	4.1secs
Mpg overall/touring	19.1/23.7
Mph/1000rpm in top	27.1
Kerb weight*	1865kg
Date tested	6.6.90

Mercedes-Benz G-Wagen 300GEL

★★★ £33,529

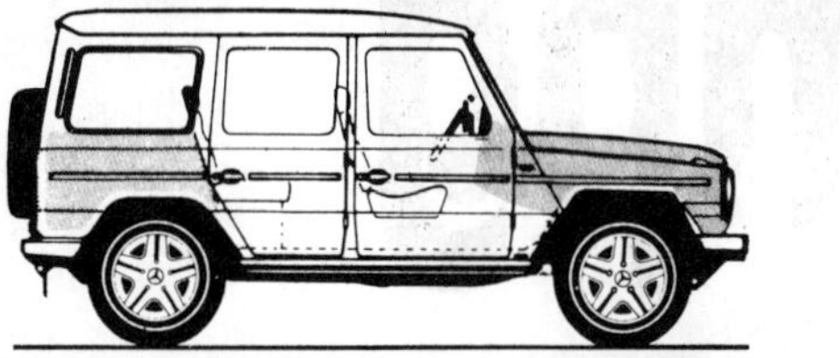

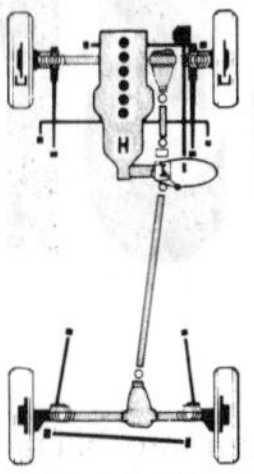

Length 4623mm **Width** 1700mm **Front track** 1425mm **Wheelbase** 2850mm
Height 1975mm **Rear track** 1425mm **Weight distribution front/rear** n/a

With a redesigned permanent four-wheel drive system and improved interior, the G-Wagen is more capable than ever. Build integrity and finish are first class, as is the car's sheer go-anywhere tenacity. Super-smooth straight-six engine is a major asset although 170bhp isn't much to propel the heaviest car here. Utilitarian Tonka Toy looks make even the Range Rover look sleek.

Capacity	2960cc
Power	170bhp/5500rpm
Torque	173lb ft/4500rpm
Max speed*	102mph
0-62mph*	13.0secs
30-70mph through gears	n/a
Standing quarter mile	n/a
30-50mph in kickdown	n/a
50-70mph in kickdown	n/a
Mpg overall/touring	n/a
Mph/1000rpm in top	n/a
Kerb weight*	2220kg
Date tested	n/a

* Manufacturers' claimed figures

Mitsubishi Shogun V6 LWB

★★★★ £22,219

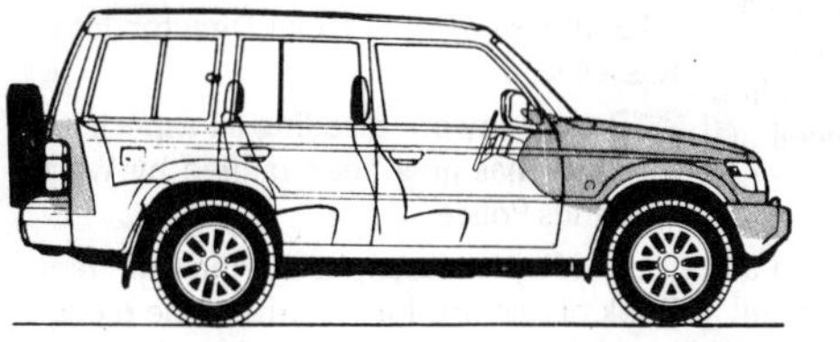

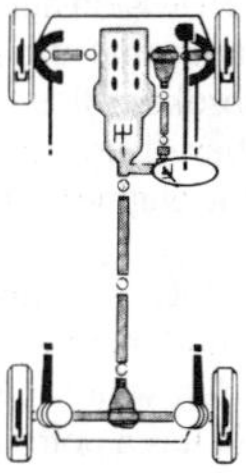

Length 4724mm **Width** 1786mm **Front track** 1486mm **Wheelbase** 2725mm
Height 1869mm **Rear track** 1486mm **Weight distribution front/rear** 46/54

The Mk II Shogun shows improvements over its forebear in nearly every area. Overall refinement is now good enough to make the V6 petrol Shogun feel like a genuine alternative to an executive car and its space, in LWB form, matches many estate cars'. 'Super Select' 4wd allows the driver to engage either rear-wheel drive or 4wd while on the move. Capable and fine value.

Capacity	2972cc
Power	147bhp/5000rpm
Torque	174lb ft/4000rpm
Max speed	104mph
0-60mph	13.1secs
30-70mph through gears	14.2secs
Standing quarter mile	19.7secs
30-50mph in fourth	9.4secs
50-70mph in top	15.0secs
Mpg overall/touring	16.1/19.2
Mph/1000rpm in top	20.9
Kerb weight	2055kg
Date tested	26.6.91

Tdi Range Rover clocks 53mpg

A STANDARD Range Rover fitted with Land Rover's own 200 Tdi diesel engine has covered the 832 miles from John O'Groats to Lands End without refuelling.

Despite strong winds, road works and hail storms, the Range Rover Vogue Tdi, introduced to the UK market this month, returned 53 mpg at an average speed of 41 mph.

The Range Rover, which has an 18 gallon fuel tank, was driven by economy driving specialist John Taylor.

He said: "I have done this journey several times, but I have never seen so much rain, wind and traffic. I am sure that had we had a clearer run the results would have been much better."

WPC Tracey Segarty of the West Midlands Police Force F Division acted as official observer ensuring that the fuel tank seal remained unbroken, that all traffic laws were strictly obeyed, no free wheeling occurred and that the engine was kept running in heavy traffic.

When the Range Rover was refuelled at Lands End it was found that there was still nearly two gallons left which means that the Range Rover Tdi has a potential range of more than 920 miles.

This puts many parts of Europe such as Oslo, Madrid and Salzburg within the range of just one tank of diesel.

Land Rover Commercial Director John Russell said:

"The paperwork detailing all aspects of the drive has now been ratified by West Midlands Police."

Land Rover has applied to the Guinness Book of Records for the drive to be recognised as an economy record for a dual purpose 4x4 vehicle."

Range Rover County LWB

New suspension, new traction control, bigger engine, longer wheelbase. And it's still chic.

BY WILLIAM JEANES

Once, a new Range Rover model meant two things: a price increase and an increase in the displacement of its V-8, an engine that can be traced to the 1963 Buick Skylark. For 1993, the Range Rover County got both, but bigger news accompanied the welcome addition of twenty horsepower.

The wheelbase on the Range Rover County has been extended a full two-thirds of a foot, to 108.0 inches. Because the rear compartment got seven of those eight inches, riders with legs perceptibly longer than those of Henri Toulouse-Lautrec can luxuriate in a spacious passenger enclosure free from the risk of patella fracture and the horrors of claustrophobia.

As welcome as the added room is (and there's no cheating—the cargo capacity with the rear seat in use is unchanged but gains six cubic feet with the seat folded), the most significant news concerns Range Rover's new Electronic Air Suspension (EAS), which can raise or lower the ride height through a span of 5.1 inches.

The previous Range Rover suspension offered the most comfortable off-road ride in the business, an honor due in large part to its all-around coil springs and long, bump-absorbing suspension travel aided by a load-leveling device. Any new suspension had to maintain the marque's tradition of off-road utility, capability, and comfort and add luster to street and highway operation. The new airborne suspension does all of this, and with some stylish flourishes.

We experienced the new Range Rover in Morocco (see "Touring," page 173) where we drove it on some tough desert roads and through some balky, boulder-strewn river beds. With no reservations, we can pronounce the off-road ride comfort as equal to or better than what was already extraordinary.

PHOTOGRAPHY BY SUSAN SMITH JEANES

the Verdict:

Highs: Comfort, capability, cachet.

Lows: Lofty price and the risk that the Duchess of York may own one.

The Verdict: Like Chateau Lafite-Rothschild, the best of its kind but you'll pay a price.

The EAS does a number of useful things in addition to its normal duties. Though it may sound laughable, given the boxy shape with which the Range Rover confronts aerodynamicists, the suspension automatically lowers the body by three-quarters of an inch—from the "Normal" suspension setting to "Low Profile"—once a speed of 50 mph has been attained. (Note that the newly bored-out 4.2-liter V-8 develops 200 horsepower and 251 pound-feet of torque, thereby getting you to 50 mph far more quickly than before.)

Off-road, should you high-center the vehicle, the air suspension automatically assumes its "Extended Profile" position, sending the wheels (and axles) downward over 2.7 inches in search of firmer footing. In the event of a flat tire, we learned that this feature will cause the injured tire to try to remain flat on the ground if the engine is not turned off before you institute jacking procedures. Another off-road EAS option is the "High Profile" setting that raises the ride height by 1.6 inches for rough going.

Last on the air suspension's trick list is the "Access" mode. No longer must one drive up to the opera house and watch as women either expose themselves or rip the seams out of tight skirts as they descend to street level from their leather thrones. The driver can now punch one of the three suspension-control buttons that reside in a small panel below and to the right of the steering column, and the vehicle will lower itself by 2.4 inches to a more accessible position for entry or exit. The front end lowers first, a thoughtful procedure that prevents the headlights from dazzling anyone in front of the vehicle at night. For the same reason, when the vehicle regains its normal ride height, the rear end rises first, not unlike a camel though with infinitely less noise and odor.

To the EAS system and the existing anti-lock braking system and full-time all-wheel drive, the 1993 version brings an excellent electronic traction-control system, the first such device to appear in a sport-utility vehicle and one with obvious off-road advantages. This festival of electronic magic means that only the truly inept dunce will get in serious off-road trouble in a Range Rover.

For 1993, Land Rover North America offers three vehicles: the new County LWB, a 100.0-inch-wheelbase County that retains the 3.9-liter V-8, and the rugged Land Rover Defender 110.

The County LWB costs $49,825 fully equipped. Land Rover's U.S. boss Charlie Hughes told us that Land Rover's vehicles are not subject to the luxury tax, because the IRS does not consider them luxury cars. Those who buy the new, lavishly equipped County LWB could dispute the IRS definition—and for good reason: the Range Rover County LWB is a luxurious, easy-riding sport-utility that continues as the gold standard of off-road performance and comfort. •

Vehicle type: front-engine, four-wheel-drive, 5-passenger, 5-door wagon

Price as tested: $49,825

Price and option breakdown: base Range Rover County LWB (includes $625 freight), $49,825

Major standard accessories: power steering, windows, seats, locks, and sunroof, A/C, cruise control, tilt steering, rear defroster and wiper

Sound system: a/d/s/ AM/FM-stereo radio/cassette/CD player, 7 speakers

ENGINE

Type .... V-8, aluminum block and heads
Bore x stroke .... 3.70 x 3.03 in, 94.0 x 77.0mm
Displacement .... 258 cu in, 4275cc
Compression ratio 9.0:1
Engine-control system .... Lucas with port fuel injection
Emissions controls .... 3-way catalytic converter, feedback fuel-air-ratio control, EGR
Valve gear .... pushrods, hydraulic lifters
Power (SAE net) .... 200 bhp @ 4850 rpm
Torque (SAE net) .... 251 lb-ft @ 3250 rpm
Redline .... 5200 rpm

DRIVETRAIN

Transmission .... 4-speed automatic with lockup torque converter
Final-drive ratio 3.54:1
Transfer-gear ratio L:3.24:1, H:1.21:1

Gear	Ratio	Mph/1000 rpm	Max. test speed L/H
I	2.48	2.9L/7.9H	15/41(5200/5200 rpm)
II	1.48	4.9L/13.3H	26/69 (5200/5200 rpm)
III	1.00	7.3L/19.6H	38/102 (5200/5200 rpm)
IV	0.73	10.0L/27.0H	52/111 (5200/4100 rpm)

DIMENSIONS AND CAPACITIES

Wheelbase .... 108.0 in
Track, F/R .... 58.5/58.5 in
Length .... 183.0 in
Width .... 71.4 in
Height .... 72.2 in
Frontal area .... 29.1 sq ft
Ground clearance .... 8.0 in
Curb weight .... 4634 lb
Weight distribution, F/R 49.5/50.5%
Fuel capacity .... 21.6 gal
Oil capacity .... 5.4 qt
Water capacity .... 12.0 qt

CHASSIS/BODY

Type .... full-length frame with rubber-isolated body
Body material .... welded steel stampings and aluminum stampings

INTERIOR

SAE volume, front seat .... 53 cu ft
rear seat .... 51 cu ft
luggage space .... 36 cu ft
Front seats .... bucket
Seat adjustments .... fore and aft, seatback angle, front height, rear height
Restraint systems, front .... manual 3-point belts
rear .... manual 3-point belts, center lap belt
General comfort .... poor fair **good** excellent
Fore-and-aft support .... poor fair **good** excellent
Lateral support .... **poor** fair good excellent

SUSPENSION

F: .... rigid axle, 2 leading arms, Panhard rod, air springs, anti-roll bar
R: ..... rigid axle, 2 trailing links, 1 central A-arm, air springs with automatic leveling, anti-roll bar

STEERING

Type .... recirculating ball, power-assisted
Turns lock-to-lock 3.4
Turning circle curb-to-curb .... 39.4 ft

BRAKES

F: .... 11.8 x 0.9-in vented disc
R: .... 11.4 x 0.5-in disc
Power assist .... vacuum with anti-lock control

WHEELS AND TIRES

Wheel size .... 7.0 x 16 in
Wheel type .... cast aluminum
Tires .... Michelin XM+S244, 205TR-16
Test inflation pressures, F/R .... 28/41 psi

CAR AND DRIVER TEST RESULTS

ACCELERATION — Seconds

Zero to 30 mph 3.4
40 mph 5.2
50 mph 7.4
60 mph 10.1
70 mph 14.0
80 mph 19.0
90 mph 26.2
100 mph 39.8
Street start, 5–60 mph 10.5
Top-gear passing time, 30–50 mph 5.2
50–70 mph 6.8
Standing ¼-mile .... 17.7 sec @ 77 mph
Top speed .... 111 mph

BRAKING

70–0 mph @ impending lockup .... 215 ft
Fade .... none **moderate** heavy

HANDLING

Roadholding, 300-ft-dia skidpad .... 0.69 g
Understeer .... minimal **moderate** excessive

COAST-DOWN MEASUREMENTS

Road horsepower @ 30 mph .... 9 hp
50 mph .... 24 hp
70 mph .... 51 hp

FUEL ECONOMY

EPA city driving .... **12 mpg**
EPA highway driving .... 15 mpg
C/D observed fuel economy .... **10 mpg**

INTERIOR SOUND LEVEL

Idle .... 53 dBA
Full-throttle acceleration .... 76 dBA
70-mph cruising .... 73 dBA
70-mph coasting .... 72 dBA

CURRENT BASE PRICE* dollars x 1000

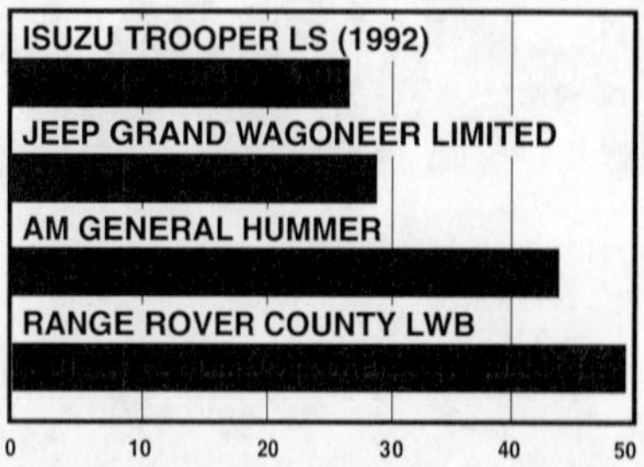

ACCELERATION seconds

■ 0–60 mph
■ 1/4-mile

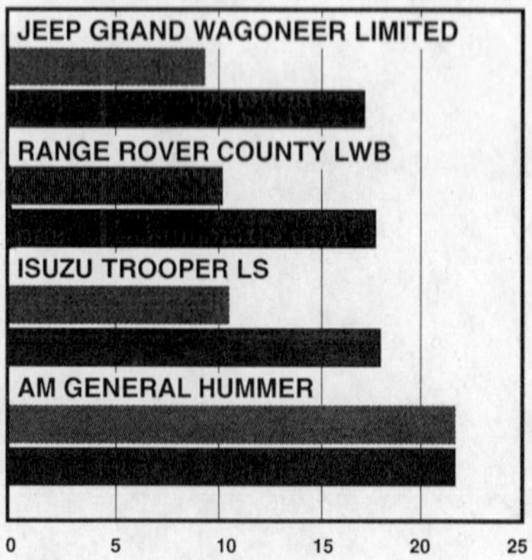

70–0 MPH BRAKING feet

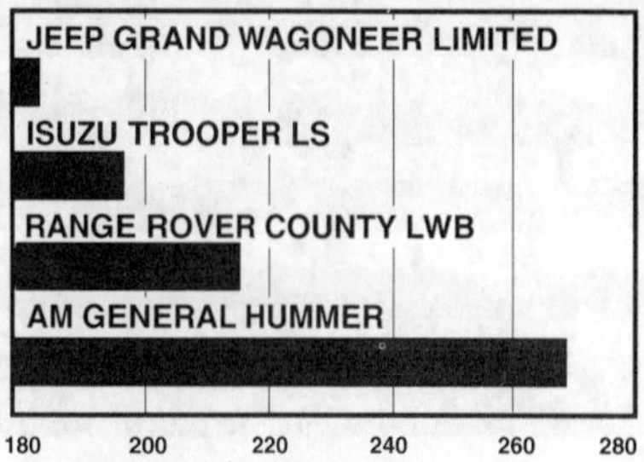

ROADHOLDING 300-foot skidpad, g

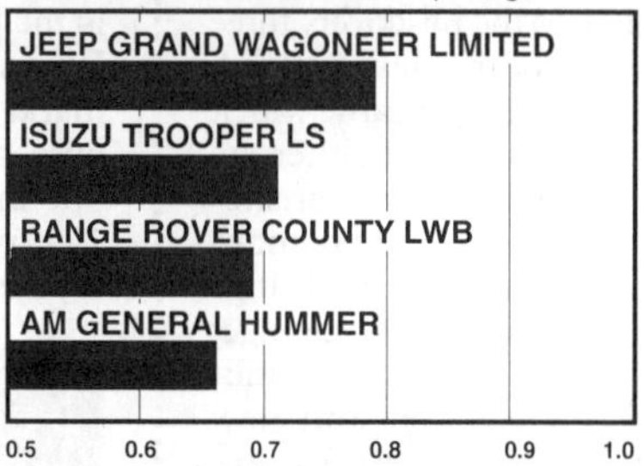

EPA ESTIMATED FUEL ECONOMY mpg

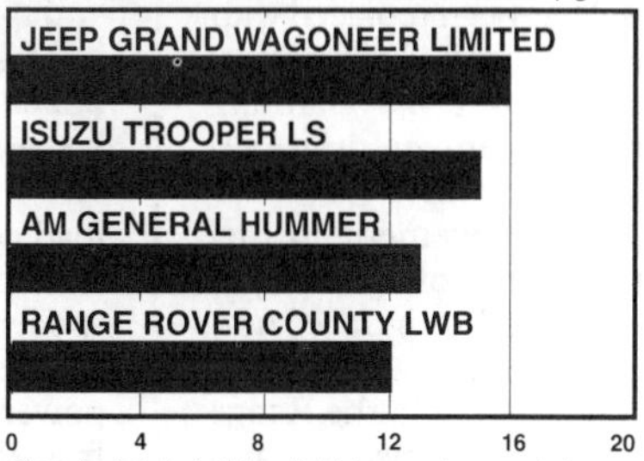

*Base price includes freight, applicable luxury and gas-guzzler taxes, and any performance options.

Long-range Rover

The Range Rover makes big gains in economy by using the Discovery's diesel engine. By Howard Lees

WHEN LAND ROVER launched the Discovery three years ago, the new 200 Tdi turbo diesel that came with it was said to have neither the right power characteristics nor sufficient refinement to find its way into the Range Rover. But times change; first it was fitted to the utilitarian Defender, and now the Range Rover's Italian-built 2.5-litre VM turbo diesel has been replaced with the Tdi unit. As a result, Land Rover is claiming better refinement, driveability and economy.

An improvement in fuel consumption of up to 30 per cent is hard to ignore. And it should be a powerful marketing tool in the Range Rover's cost-conscious and largely diesel-led European markets.

Just one model is available in the UK — the Vogue Tdi. Its list price of £27,899 marks a £1500 price hike over the outgoing version but includes all the other 1993 model year refinements, such as the better stereo, new load cover and, for the first time on the diesel, a front spoiler and driving lamps.

Better economy with Tdi

The direct injection Tdi powerplant is unchanged from its turbocharged and intercooled Discovery guise, with 111bhp at 4000rpm and 195lb ft of torque at just 1800rpm. The indirect injection VM powerplant mustered 119bhp at 4200 rpm and 209lb ft of torque at 1950rpm. Gearbox, transfer and final drive ratios are unchanged. Both engines have aluminium cylinder heads and cast iron blocks, but the Tdi unit has its roots in the Land Rover's original diesel engine of 30 years ago and is bulkier and heavier. At a claimed kerb weight of 2052kg, the Range Rover Vogue has gained 37kg.

Tdi is quieter inside

None of which is destined to improve performance. With a top speed of 94mph and a 0-60mph time of 16.6secs, the Tdi is 1mph and 0.8sec slower than its predecessor. Continental buyers in particular won't worry about that when they look at the fuel consumption figures: 41.0mpg, 29.1mpg and 32.9mpg at 56mph, 75mph and on the urban cycle respectively make the Tdi up to 30 per cent better than its Turbo D predecessor. A fuel capacity of 81.8 litres (18 gallons) gives the Range Rover Tdi a 500-mile range between fill-ups.

As well as extra insulation of the engine bulkhead, floor and transmission tunnel introduced across the Range Rover range with the launch of the air-sprung LSE, the

Tdi engine gains a two-part composite enclosure around the bellhousing, a rocker cover in a similar material and rubber/stainless steel damping pads inside the sump to tackle vibration. The result, according to Land Rover, is a reduction in interior noise of five decibels at 3000rpm.

It also boasts that the Tdi has cut smoke and clatter on cold start-up, to which the VM was prone. At tickover the Tdi does seem quieter

Fewer gearshifts needed

inside, although passers-by would be hard pushed to detect an improvement.

Certainly, the new engine has stronger bottom-end torque and a much smoother build-up of boost pressure than the VM unit. All of which means the Range Rover is easier to drive and requires less gearshifting than the old diesel. Only up long motorway inclines do the extra weight and reduced top-end power conspire to

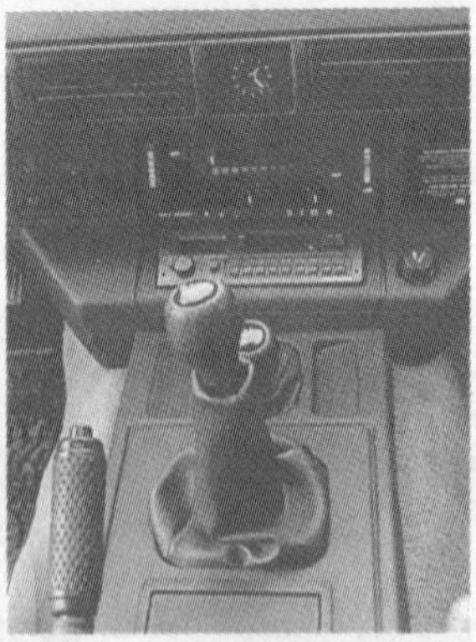

Minor changes to cabin

limit the Tdi's urge in fifth.

Up to 3000rpm — around 75mph on the clock — the Tdi is acceptably smooth and noticeably more refined than the VM, but it is by no means impossible to tell you are in an oil-burner. In fact, the noise from the direct injection engine, though it may be at lower volume than before, is noticeably coarser, especially as you wind the unit up towards 4000rpm.

The anti-roll bars introduced a couple of years ago mean the Tdi handles tidily and with considerably less roll than of old. But, having sampled the on and off-road isolation provided by the new EAS air suspension, you can't help feeling a little cheated by mere steel coils. Land Rover says there is no engineering reason why this system can't be fitted to the Tdi but won't offer it unless there is significant demand.

What you get with the Range Rover Tdi is not so

Flexible Tdi diesel engine means improved off-road performance for Range Rover

Factfile

Range Rover Tdi

How fast

0-60mph	16.6secs
Top speed	94.0mph
MPG: urban	32.9
56mph	41.0
75mph	29.1

All manufacturer's claimed figures

How much?
£27,889 **On sale in UK** now

How big?

Length	4447mm (175ins)
Width	1813mm (72ins)
Height	1792mm (71ins)
Wheelbase	2540mm (100ins)
Weight (claimed)	2052kg (4525lb)
Fuel tank	81.8 litres (18 galls)

Engine
Max power 113bhp/4000rpm **Max torque** 195lb ft/1800rpm **Installation** longitudinal, front, four-wheel drive **Capacity** 2495cc, 4 cyls in line **Made of** aluminium head, iron block **Bore/stroke** 90.5mm/97mm **Compression ratio** 19.5:1 **Valves** 2 per cyl, ohv

Gearbox
Type 5-speed manual **Ratios 1st** 3.69 **2nd** 2.13 **3rd** 1.40 **4th** 1.00 **5th** 0.77 **Final drive ratio** 3.54 **Transfer box ratio** 3.24 (high) 1.21 (low)

Suspension
Front live axle, coils and dampers **Rear** live axle, coils and dampers with self-levelling

Steering
Type worm and roller

Brakes
Front vented discs **Rear** discs **Anti-lock** standard

Wheels and tyres
Size 7Jx16ins **Made of** alloy **Tyres** Michelin 205 R16 XM

Made by
Land Rover, Lode Lane, Solihull, West Midlands B92 8NW. Tel 021 722 2424

much an improvement in refinement, although Land Rover has done its homework on noise, vibration and harshness. Rather, it represents a quantum leap in fuel economy. This is the direct benefit of using one of the most fuel-efficient automotive engines yet built.

It's an honest installation, with much better flexibility that improves off-road performance rather more than it does on the road. But it would have been nice to see some the cost savings of using an in-house engine passed on to the customer. ■

★ Howard Lees, our former technical editor, died on 17 November. His wife Hilkka has asked us to run all his outstanding stories, a wish we are only too happy to honour

Δ *Range Rover Tdi now has front spoiler and driving lights as standard (photo: Richard Thomas)*

NEW DIESEL

Range Rover Tdi

THE NEW diesel Range Rover, fitted with the highly-acclaimed 200 Tdi power unit, was launched in November.

The direct injection turbo-charged 2.5-litre engine, designed and built by Land Rover engineers at Solihull, replaces the existing VM unit. The result is that fuel economy is improved by 25 per cent on the constant 56mph cycle, returning 41mpg. And the new engine also improves fuel economy on around-town driving (by 30 per cent) and on high speed cruising (by over 20 per cent).

According to official figures the Range Rover Tdi is capable of travelling over 700 miles on a full tank of fuel.

The new diesel Range Rover is claimed to have a quieter cold start, smoother town driving, and better open-road cruising. The improved driveability is the result of a better torque spread with maximum power available lower in the rev range.

The Tdi engine, first introduced in the Land Rover Discovery, produces 111bhp (83kW) at 4000rpm. Maximum torque is 195lb.ft (265Nm) at 1800rpm.

A host of refinement improvements result in a much quieter vehicle. The bell housing has a new acoustic jacket, a new rocker cover acoustic pad is fitted, and the sump also gets damping pads.

Improvements are also made to the in-car entertainment system, stowage, rear seat rake, loadspace cover, and electric windows.

The new alarm and remote locking system is also standard on the Tdi diesel Range Rover. The price in the UK will be £27,899.

The engine

The four-cylinder in-line intercooled and turbo charged high speed direct injection diesel engine produces111bhp at 4000rpm and its maximum torque is 195lb.ft at 1800rpm. The excellent bottom end torque results in outstanding towing capability.

The cylinder head, manufactured from aluminium, represents the latest direct injection technology combining light weight with excellent heat transfer properties.

The key to the engine's combination of economy, performance, and clean combustion is the high precision of "swirl port" design, and the specification of two-spring injectors.

Fuel consumption is 32.9mpg (8.57 1/100km) on the simulated urban cycle; 41.0mpg (6.88 1/100km) at a constant 56mph; and 29.1mpg (9.69 1/100km) at a constant 75mph.

The 200 Tdi, originally launched in the Land Rover Discovery, has been modified to reduce engine noise. A moulded rocker cover is added to cut noise levels from the top end of the engine, and constrained layer damping pads are applied to the sump. In addition a bell housing acoustic blanket is fitted.

Insulation

A comprehensive package of improvements has been introduced to make the diesel Range Rovers quieter inside. Additional noise absorbing material has been put in the passenger side bulkhead, pedal box covers, tunnel and front floor, seat box covers, intermediate floor, body mountings, and extractor vent cavities.

The overall effect of noise insulation measures to the body and engine is to reduce interior noise from 75dbA to 70dbA at 3000rpm.

Spoiler

The front spoiler incorporating driving lamps, currently fitted to petrol Range Rovers, is now fitted as standard on the Range Rover vogue Tdi.

Δ *Suitably badged*

The 200 Tdi engine in the Range Rover. Special attention has been paid to sound insulation and the elimination of vibration into the passenger compartment ▷

Engine	2.5-litre turbo charged direct injection diesel		
Cylinder Head Material	Aluminium		
Cylinder Block Material	Cast iron		
Number of Cylinders	4		
Cylinder Layout	In-line		
Bore	90.47mm		
Stroke	97.00mm		
Capacity	2495cc		
Valve Gear	ohv		
Compression Ratio	19.5:1		
Fuel Injection	Bosch VE4/11F		
Max power output	111.3bhp (83kW) at 4000rpm		
Max torque	195lb.ft (265Nm) at 1800rpm		
Transmission			
Manual	mph/1000rpm at gearbox input		
Gear	Ratio	High	Low
5th	0.770:1	25.45	9.46
4th	1.000:1	19.59	7.28
3rd	1.397:1	14.03	5.21
2nd	2.132:1	9.19	3.41
1st	3.692:1	5.31	1.97
Reverse	3.429:1	5.71	2.12
Final Drive Ratio	3.540:1		
Transfer Box Ratio		2060:1	3.2444:1
Standard Suspension			
Front	Springs		Coil 150lb/in
	Dampers		Telescopic
	Anti-roll bar thickness		24mm
Rear	Springs		Coil 150lb/in
	Dampers		Telescopic
	Self Levelling		Standard
	Anti-roll bar thickness		18.5mm
Fuel			
	Tank capacity	18 gallons (81.8 litres)	
	Fuel grade	Diesel	
Fuel Consumption	mpg	L/100km	
Urban cycle	32.9	8.57	
Constant mph	41.0	6.88	

Constant 75mph	29.1	6.69	
Steering			
Type	Worm and roller (Adwest)		
PAS	Standard		
Brakes			
Front	11.73in. dia (298mnm) ventilated discs		
Rear	11.42in. dia. (290mm) plain		
Servo	Standard		
ABS	WABCO: Optional		
Handbrake operation	Transmission brake drum on rear output from transfer box		
Weights			
Max gross weight	2510kg (5522lb)		
EEC kerb weight	2052kg (4525lb)		
Towing weight	Fully braked	4000kg	
	Over-run brakes	3500kg	
	Without brakes	750kg	
Wheels			
Construction	Alloy		Steel
Rim width	7J		6JK
Tyres	Michelin 205 R16 XM+S 244 tubeless		Michelin 205 R16 XM+S tubed
Dimensions			
Wheelbase	100.0in	2540mm	
Overall length	175.0in	4447mm	
Track front	58.5in	1486mm	
Track rear	58.5in	1486mm	
Height	71.0in	1792mm	
Width	72.0in	1813mm	
Ground clearance	7.5in	190mm	
Luggage capacity (seats up)	36.2cu.ft	1.02cu.m	
Luggage capacity (seats down)	70.8cu.ft	2.00cu.m	
Performance			
0-60mph	16.6 secs		
Maximum speed	94.0mph		
Service Data			
Main service intervals	6000 miles		

TECHNOLOGICALLY ADVANCED, BUT DATED ALL THE SAME. THE RANGE ROVER HAS ALWAYS BEEN A MYSTERIOUS COMBINATION

AIR OF UNCERTAINTY

Report and pictures by test editor Ian Shaw

LOVE IT or hate it, the Range Rover is a legend in its own lifetime. At its launch in 1970, it created a whole new class of vehicle not seen before, with the off-road ability of a Land Rover and the on-road comfort of a Rover saloon. At one point the waiting list for buyers ran into years, not weeks or months.

It could not sustain the position forever, and was gradually moved upmarket into what became known as a 'luxury off-roader'. The latest version, the LSE, is the most significant model to carry the name since that first embodiment 23 years ago.

The development of the model has been steady; indeed, its pace has quickened of late as more and more competitors have come on the scene. It spent the first 12 years of its life as a two-door, carb-fed, four-speed manual gearbox machine. The first four-door model, and option of three-speed auto appeared in 1982. The five-speed manual arrived a year later, with electronic fuel injection and four-speed ZF auto box becoming available in 1985.

After just 16 years, the diesel version was unvailed, and the next milestone was a viscous-coupled centre diff in 1989. In 1990, Land Rover celebrated its flagship's 20th birthday with a 3.9-litre version of the V8 and ABS brakes.

The LSE would appear to be the final chapter of this particular model's life, save for the replacement of the VM turbo-diesel with Land Rover's own Tdi unit (**IOR**, *Frontline News, Jan*). A entirely new machine has been on the cards for a while – the LSE introduces many of its features.

OVERVIEW

Some areas of the new model are familiar Range Rover territory. Ladder-frame chassis, live axles, 16in wheels with 205 section rubber, the V8 engine and the overall shape. However, a few very notable changes are to be seen. The wheelbase is 108in on the LSE, 8in more than standard, the V8 engine now displaces the best part of 4.3-litres and produces 200bhp and 250lb/ft of torque.

Transmission is four-speed auto main gearbox with two-speed transfer giving the usual permanent four-wheel-drive via a viscous-coupled centre differential. But just above the live axles, it's all change. The coil springs have gone, replaced by compressed air suspension giving five different height settings. These comprise two driver-selected modes and three automatic ones. The vehicle's standard ride height is the same as the coil-sprung models; at more than 50mph, the vehicle adopts 'low profile' dropping 20mm from 'standard'. Then come the two driver-selected modes: 'high profile', 40mm above 'standard' for off-road use, and 'access mode' where the vehicle is dropped 60mm below 'standard' for easier access.

This leaves just the third of the auto modes ('standard' is an automatically selected mode when travelling at up to 50mph), which is 'extended profile'. When the vehicle has grounded its underside, presumably sensed by the system detecting loss of load on the suspension units, the 'stroke' of the suspension is increased to maximum to regain clearance, and hence traction. It has a limit to what it can achieve, but it's a nice idea. Basic suspension geometry is unchanged.

As a by-product of the ABS system, the LSE is equipped with electronic traction control. This uses the ABS wheelspeed sensors, on the rear axle only, to detect one rear wheel spinning, as the diff pushes its torque to the least resistance, and under electronic control, brakes the spinning wheel, to balance out the resistance and achieve more even traction. A manual version has been available for years: fiddle brakes, something every sporting car or modified 4x4 vehicle trials driver will be familiar with.

The LSE's extra length is between 'B' and 'C' pillars. The driving position and standard luggage space are unchanged. The additional 8in is in the rear passenger footwells – and in the luggage space, once the rear seats are folded.

BEHIND THE WHEEL

So, in principle, the LSE is a pretty up-to-the-minute machine; however, once inside, it's something of a time capsule.

Top: Air suspension gives an improvement of just 2 degrees to approach and departure angles over 'standard' ride height. Centre: Articulation is all you'd expect, and the LSE made short work of our cross-axle gulley. Bottom: Ramp-over is improved by one degree over 100in coil-sprung model

The basic interior layout hasn't changed in 10 years, and it shows. At the time it was designed, the facia would have been fine, with very little in the way of electrical goodies requiring switching. Not any more. Major functions, lighting and wash/wipe for both front and rear screens, are taken care of by two full-size column stalks and two smaller toggles on the column. Everything else – rear fog, heaters for front and rear screens, front driving lamps, cruise control master, interior light override – is positioned low and out of direct sight just to the left of the driver's left knee.

Add to this the even lower-mounted ICE unit and seat heater switches as well as the controls for the air suspension, and you can see why it's an ergonomic disaster. The upper facia is an empty expanse. It would be bad for a £20,000 vehicle, but for almost 40 grand, it's an absolute joke. The electric window switches are aft of the gear selector on the sloping front face of the cubby box and the control for the electric mirrors is the only well-positioned switch in the shop.

If that were not enough, the instrumentation is basic. Just speedo, tacho, fuel and temperature gauges.

Visibility is fine at the front. The squared-off '70s styling works with you here, the corners of the front wings are easy to sight, and for a big vehicle, the LSE is easy to place. Rearward visibility could be better in the rain: the rear wiper sweeps an area equivalent to your average cat-flap.

The leather seats are electrically adjusted, with a memory setting for the driver's. They look good, but lack lumbar support, have only token lateral support and the cushions are as short now as they've always been. It is from here more than anywhere else that the LSE feels old. The steering wheel is also too low, and unbelievably on a vehicle of this price, is non-adjustable.

ON THE ROAD

A bank of warning lights greets you as you turn the key for the first time. The ABS and ETC (Electronic Traction Control) self-check, the air suspension shows all of its positions simultaneously as a circuit and bulb check, all in addition to the usual engine and brake fluid check lights.

The V8 starts a little reluctantly when cold, it doesn't cough or run roughly but cranking is a somewhat protracted affair and in the first few yards it sounds 'tappety' and rather under fuelled, this could, however, have been a characteristic of our particular example.

Once underway, the V8 is effortless and the ZF auto box a lot smoother than it used to be, the Range Rover finally rivalling Jaguar and Mercedes-Benz as a refined road machine. The performance, however, is no match for a 4-litre XJ6 or 300 coupe. The Vogue LSE's version of the Rover V8 produces 200bhp but that translates into just

93bhp/tonnes, par for the course for most mid-hatch or estate cars.

The 0-60mph time of 9.9 seconds is more than reasonable for a vehicle of this size and obvious aerodynamic failings. All out, it's good for 112. It may be no match for the cars by which potential LSE buyers may be tempted, but it's still the fastest of the full-size off-roaders.

The torque is pretty obvious too, all 250lb/ft of it, even if it is developed at a rather high 3,250rpm. The auto slips into second almost immediately on a light throttle, it's in third by around 20mph and will slip into top at little more than 30mph. The torque coverter lock-up doesn't come in until the top side of 50mph, however.

With around 28mph indicated for every 1,000rpm, it makes for a relaxed cruiser. At the legal limit, the engine is turning over at just 2,600rpm, although it is worth remembering that peak torque isn't generated until 87mph, which I suspect is just a touch on the fast side for towing!

Trying harder in the lower gears is made easy by the sensitive kick-down and the ease with which the box can be flicked between 3rd and top, particularly around twisting back waters where the thing can be left in 3rd to prevent the box shuttling back and forth too frequently, bringing all the torque at a little over 40mph and the ability to nudge the legal limit at peak power revs. If your pocket can stand the fuel bills the Vogue LSE is as quick as you're likely to need.

If you're expecting the air suspension to transform the vehicle's road manners then you'll be rather disappointed. It feels just like the 100in wheelbase coil-sprung Vogue. Body roll is well controlled these days, the LSE has anti-roll bars just like its smaller brother and was much better controlled in this respect than the Discovery V8i auto which was in our hands shortly before.

Through long, fast corners it can be made to hold on quite well. The steering isn't particularly communicative, it never has been, but you have a fair idea as to just how much the front wheels are understeering. It's not excessive, but shut the throttle and even with the lesser engine braking of the automatic there is still a distinct change in attitude from a rather tighter-than-neutral, to full-blown roll-oversteer depending on the conditions of the corner and the road speed.

Through tighter corners, the understeer is more pronounced if you leave the LSE to its own devices; it's safe and predictable but not particularly rewarding. However, you can make use of its roll-y character by coming off the power quite late into the corner, whereupon the LSE will swing in a little, the understeer is banished, and the steering feels as if it's attacking the curve in a much more effective manner. As you plant the throttle to the floor it peels away from the apex in a nice, stable neutral stance.

Nothing changes: learn the Range Rover's foibles and use them to your advantage; drive as you would a car and it will disappoint.

Something the air suspension hasn't cured, not that it is in its power to do so, is the wavering and wandering at cruising speeds over undulating 'A' roads. There is simply too much compliance, it feels remote. Although it is quite secure if you drive it every inch of the way, it feels rather sloppy and will hardly impress those with mid-range-Merc money.

The brakes are strange, to say the least. The ABS is simply fantastic. This technology has come a long way in a short time. I remember driving early systems from several manufacturers where the thing pumped the brakes on and off at a slower rate than most competant rally drivers can achieve, treating all wheels equally without multi-channels.

Lately, however, things have taken off, and the modern ABS system using various 'select' logic readings can cope with a multitude of situations. This one's no different. We tried it under conditions of merely damp, to streaming wet, tarmac and gravel. It worked well, not only providing high degrees of retardation but successfully allowed the steering to function – ABS's main claim to fame – and appeared to be 'intelligent' enough to allow unladen wheels to lock momentarily in the interests of braking forces on those with grip.

So, why are the brakes strange? Simply because they have the worst feel of any vehicle over £10,000 I've come across. It may simply be an inability on the designer's part, it may be looked upon as unimportant: 'Who needs to feel the braking force, we've given them ABS, they're safe.'

I couldn't disagree more, the pedal feels like a block of hardwood operating the brakes via a micro-switch. On or off. That's how it is, unprogressive, unresponsive; unforgivable.

What everyone wants to know about, in this vehicle with the most advanced suspension of any off-roader, is the ride quality.

The simple fact is the LSE rides very well. Just like the coil-sprung models always did. That's it, I'm afraid. Under all the test conditions I couldn't find anywhere that the air sprung version was noticeably better than the coil-suspended machine. The comparision was easy, we had just handed back a Discovery V8i, it rode just as well, it rolled more and it had a little more braking dive, but over merely indifferent patchy urban roadscapes, it was no worse.

The Vogue LSE has all the lateral uneasiness of previous Range Rovers. It wallows almost as much, the anti-roll bars – standard for the past couple of years – and extra damping effect provided by the longer wheelbase and slightly greater kerb weight seem to have as much to do with the ride quality as the new suspension.

Potholes comprehensively catch it out, the resulting crash being a match for that of any MacPherson strut-suspended five-door family box. The ride is good, make no mistake, but be equally aware, it always has been. The LSE has broken no new ground.

OFF THE ROAD

The air suspension is again the first port of call here. The 'high-profile' mode, which lifts the chassis another 40mm above the axles compared with standard 'profile' on the electronic air suspension or coil-sprung models, gives an improvement of just two degrees to the

Above: Interior is plush, but ergonomically poor. Right: Rear seat passengers get all the additional space

approach and departure angles. It seems to have been initiated to try to offset the disadvantage in ramp breakover angle which the extra 8in of wheelbase produces. As it is, the ramp angle is improved by one degree over that of a coil-sprung 100in and is two degrees behind an air-sprung 100in vehicle. Every little helps – I'm always happy to see an improvement in a vehicle's off-road ability no matter how slight.

Further to this goal, the computer controlling the suspension will go into 'extended profile' if the vehicle should belly-out, lifting the chassis another 30mm over 'high profile' to regain traction. That's all very well, but seldom is a vehicle halted by bellying-out, at least not in over-populated over-used British off-road conditons. Here, it is more likely to be brought to a halt by grounding its axles in ruts created by larger-shod vehicles.

The LSE, still on 205R16 tyres and the same Range Rover axles has the 'worst-in-class' minimum ground clearance of 7.5in that it did 23 years ago.

Articulation is all you'd expect from this vehicle. However, we suspected that it wasn't quite as supple in 'high profile' as the coil-sprung version. With anti-roll bars limiting the drop of the unladen wheel to some extent, we were looking at a lot of variables at once. Nevertheless, it made short work of our cross-axle test and even appeared to sit more level than most vehicles.

The vehicle is just as effortless off the road as it is on. The big V8 has plenty of torque at the bottom end of its rev range although I did notice I had to use considerably more throttle in places where I would not have expected to. The peak of 3,250rpm is rather high even if 250lb/ft is developed. Generally, however, it is the gearbox, not the engine which impresses.

Once under the influence of the deep low ratio, the four-speed auto is in its element. It permits the merest trickle of power to the wheels, it allows the vehicle to crawl over rocks and through ruts which would send a manual version bouncing, since even the road speed in low-1st at idle would be considerably higher than the velocities we were using in places. At the other end of the scale, the torque converter provided the LSE with impressive acceleration when a little momentum was required with too little space in which to generate it.

Engine braking was predictably inferior to a manual, but in most conditions not to any real detriment to performance or the vehicle's body panels, thankfully! I took to a little left-foot braking under power to provide the balance between grip and retardation on the steeper parts and over some of the larger rocks.

One thing which disappointed was the ETC (Electronic Traction Control). This uses the ABS sensors to apply the brakes to whichever rear wheel is spinning in any given situation up to a speed of 31mph. The idea is to transfer the wasted torque from the spinning wheel, by resisting it with the brakes and pushing it to its opposite number at the other side of the differential.

The facia-mounted warning light told us that it was operating but the only reaction at the rear wheels was a shuttling to and fro of the applied engine torque from one wheel to the other as they both spun and snatched intermittently. It felt for all the world like an old Powr-lok diff and was powerless to assist the LSE in climbing a fairly innocent-looking ridge at a slightly cross-axled position. Moreover, the handbook warns that should the tell-tale flash, the unit has switched off owing to overheating after the 'unlikely' event of more than a minute of continuous use. I know of several climbs which require all the traction you can muster for well over a minute.

The company tells us that it has taken over three years of development along with WABCO. I can't help feeling that a quick nip down the road to FF Developments at Coventry for a ready-developed viscous coupling might have been a better bet. Particularly in view of such a device having replaced the centre difflock around four years ago, making the Range Rover more 'layman friendly' and ironically paving the way for successful operation of ABS brakes.

The ride off road is as good as they come. The lurching which afflicts Range Rovers over cross-ruts and rock-strewn tracks is still there but seems to very well damped these days. None of this is probably of much interest to potential buyers: there are hundreds of stories told at Land Rover club dinners around Britain of the bloke who owned 15 Range Rovers in 17 years and never locked the diff or engaged low box on any of them.

More than any other, the Range Rover has always been the on-road off-roader. While this model doesn't shy away from the off-road side of things, indeed the air suspension brings some real benefits, I can't see many getting the paintwork dirty.

ACCOMMODATION

I suddenly get the feeling that this test has told a story of the inherent contradictions which the LSE gives rise to. The interior is no different.

That extra 8in in length has all gone into the rear seat footwells. The legroom in the rear is simply massive. The standard luggage space is unchanged, the driving position still as narrow and 'short-windowed' as it has been since the first of the four-door models appeared in 1982.

The leather seats are well padded

and look great. They are, however, no match for a luxury saloon of the same price. The fronts are heated, very effectively, and this is something very welcome on a cold morning, when the leather is like ice.

The heater itself is good once the engine is up to full operating temperature. the graduation could be a little finer and the fan is noisy. However, it has air-con recirculating facility and good distribution, although the demisting, much improved over earlier models, is still less than wonderful.

Oddments space could be better. The slim door bins are as good as impossible to access with the door closed, the centre tunnel cubby is no substitute for a 'proper' glove box and despite a fairly high-spec in-car entertainment unit being fitted there is no particular space for cassettes. The front seats do, however, have pockets in their backs, which are always useful.

The ICE itself is good. An RDS with traffic information facility, which has six speakers in all, four co-axial and two bass. As an option a six-disc CD player is also available. Luggage space is unchanged.

CONCLUSIONS

Who's going to buy the LSE? Its a far more complex question than you might first think. Because so few are bought with off-roading in mind, other off-roaders hardly come into it, if they did the Mercedes G-wagen and Toyota LandCruiser VX are right in there, the Toyota with a considerable saving. But it is to the road car world that we must look for alternatives, that's where buyers look, and that's where Land Rover keeps telling us its flagship competes.

Yes, you could buy a Porsche 968 Cabriolet, or Alpine A610, both of which have performance and handling on another planet compared with the LSE. But to be fair, the LSE has to compete with saloons.

Again, the 4-litre Jaguar or Mercedes 300 CE trounce it, as does the Merc estate. The illustration is, that anyone with 40 grand to spend on a motor vehicle will have some hard thinking to do; unfortunately, Land Rover is still helping the thought process.

The build quality is appalling: it would be bad for a vehicle at half the price. The panel gaps are massive, and on our example the bonnet was out of line by about a quarter of an inch, there was painted metal visible in the interior where the carpet didn't fit properly, wires showing, some of the switches didn't operate very convincingly, there was enough play on the auto-box selector lever to show two ratios at once and the civilised aura was spoiled by a multitude of plasticky squeaks and rattles.

The LSE has furthered the Range Rover – of that you can be certain. A lot of the technology present here has been introduced prior to fitting in the all-new bodied, much-rumoured 4.5-litre V8-powered replacement for sometime when. It's needed now, because the Range Rover is, not surprisingly, showing its age.

Where Range Rovers are traditionally good, the LSE is even better: performance, ride, off-road ability. Where they are bad the LSE is just as bad: build quality, driving position, ergonomics.

It all comes down to price, and each to his own, but surely there are better ways to blow 40 grand.

Price as tested £38,395

ENGINE
V8 water-cooled electronic fuel-injected petrol with catalytic converter **Capacity** 4,287cc **Bore/stroke** 94mm/77mm **Compression ratio** 8.94:1 **Maximum output** 200bhp at 4,850rpm **Maximum torque** 250lb/ft at 3,250rpm

TRANSMISSION
Four-speed automatic main gearbox driving two-speed transfer case providing permanent four-wheel-drive with viscous-coupled limited slip

SUSPENSION
Live axles on variable rate air springs, telescopic dampers and anti-roll bars

BRAKES
Front 11.73in vented discs **Rear** 11.42in plain discs. Servo assisted with WABCO ABS

WHEELS/TYRES
7J alloy shod with 205R16

DIMENSIONS
Length 183in, (4,648mm) **Width** 71in (1,810mm) **Height** 72i, (1,853mm) **Wheelbase** 108in (2,743mm) **Track f/r** 58.5i1 (1,486mm) **Ground clearance** 7.5in (190mm)

WEIGHTS
Kerb 2,150kg **GVW** 2,620kg **Towing** fully braked 4,000kg, **over-run braked** 3,500kg **unbraked** 750kg

PERFORMANCE
(manufacturer's figures)
Maximum speed 111.8mph **0-60mph** 9.9 seconds

FUEL CONSUMPTION
Govt figures **Urban** 13.1mpg **56mph** 24.2mpg **75mph** 18.7mpg **On test** 11.9mpg

Below left: Double exposure shows the difference – just 20mm – between 'standard' and 'low-profile' ride heights. Below: The big V8 has plenty of bottom-end torque although the peak 250lb/ft arrives at a rather high 3,250rpm.

The exquisitely beautiful village of San Gimignano in northern Italy.

Exploring the mule tracks of Italy

Toby Savage takes a working trip to northern Italy in a Range Rover Tdi, and finds it the envy of Europe's motorists.

Gazing out over a landscape unchanged over the last 800 years, from the vantage point of a 12th century castle, I sat down on the tailgate of the Range Rover Tdi, poured myself a cool beer from the Koolatron 12v fridge in the back and soaked up the warmth of the late afternoon sun that was bathing the Rascino plateau in its golden light. Way below my elevated position a shepherd tended his flock of sheep, large animals, their wool a rich brown colour. The sheepdogs, similar to Golden Retrievers in appearance, but longer in the leg, fussed around the flock. If there was a paradise here on Earth this came pretty close.

I was in the Cicolano Mountains just 100 kms north of Rome as guest photographer with a team from the School of Archaeological Studies at Leicester University. It was a far cry from my usual work as an advertising photographer, having to make mundane items such as vending machines look attractive.

My trip had started the previous Friday when, after work, I loaded up the back of the Tdi with kit, suitcase, and sandwiches, and studied my map of Europe to see the best way of getting from Leicester to Italy in one long haul. Having driven down the French A6 Autoroute du soleil a year before, via Paris to Provence and found the experience harassing due to the volume of traffic, I decided instead to head across northern France from Calais, to Reims, Metz and Strasbourg, before hopping over the border into Germany, then south on the free Autobahn to Switzerland, crossing the border at Basel.

I remembered there was a motorway toll to pay on entering Switzerland so a few Swiss Francs would be needed as well as French francs for the Autoroute tolls and, of course, Italian Lire once I was in Italy. It is possible to pay French Autoroute tolls with Visa or Access or buy one of the new Telecarte cards which allows you to avoid the queues. Diesel can be purchased across Europe with credit cards, but not at every garage in Italy, particularly the smaller village ones, where cash is the only form of currency.

Wallet bulging with English cash to change on the ferry, I fired up the diesel engine and headed south to Dover to catch the 11.30pm ferry. I had not driven a Tdi Range Rover before and was interested to see how it would perform in comparison with our own 5 year old Vogue S.E. 3.5 automatic.

Instantly noticeably was the lack of burble from the rear, to be replaced by a very purposeful diesel exhaust note. Acceleration through the 5 speed manual box, though not fierce, was not noticeable slower than the V8 and I was soon cruising effortlessly down the M1 listening to the excellent optional CD player. Not having a CD player myself I had dashed to a friend's house earlier that day and robbed him of a dozen CDs, which I calculated should see me right through to Italy without a repeat. The noise of the diesel engine does intrude a little at town speeds, but is lost at speed, helped by having Pavaroti on at a impressive volume.

After a brisk run to Dover in the Vogue S.E. I would be topping up with almost a tankful of super unleaded at the slightly cheaper

The author succumbing to the temptation of a drive through the forest.

English rate before crossing the Channel. Not so the Tdi. I'd barely used half a tankful and, in fact, it was not until the far side of Metz after 435 miles that the red light came on and I refuelled with 15 gallons, at £2 per gallon, showing an average of nearly 30 mpg.

This was not intended as an economy run to try and beat John Taylor's staggering 53 mpg (LRO Jan 1993). This was real motoring, cruising at 80-85 mph with the occasional burst to 95 to get past slower traffic.

A quick wash and brush up beside Lake Como.

Archaeology students excavate a 12th Century farmhouse on the Rascino plateau watched by Paul Beavit.

It was noticeable how much respect the French and the Germans have for our very English Range Rover. Here I was in a British built product that was the envy of every European in his BMW or Mercedes. I could see them admiring the Range Rover's solid stance on the road, running their eyes over the classic shape of the 25 year old body design, oft imitated, but never improved on. It felt good and I was proud to be in it, not just for myself, but for all the staff back at Solihull who built this, the world's best all-rounder. A true Ian Botham of a car equally at home in the fast lane of the Autoroute or the mule tracks up in the mountains that I was to be working in later.

I crossed into Switzerland mid-morning on Saturday and paid the fixed 30 Swiss francs (£14) toll for unrestricted use of their Autoroutes for the next year. Useful! Halfway down through Switzerland it was time to fill up again, which was bad planing as I knew fuel would be more expensive in Switzerland, everything is. Another 15 gallons went in at £2.50 per gallon. I cursed at my inefficiency, Germany would have been cheaper and a tankful could have easily got me across to Italy, where the diesel costs about £2.40 per gallon.

By 2.30 Saturday afternoon I was dipping my toes in Lake Como having covered over 800 miles in about 19 hours. After an invigorating espresso in a lakeside cafe I headed up the peninsula that reaches into the middle of Lake Como to visit Belagio, which I had read was the jewel in the crown.

The road hugged the cliff edge with a steep drop into the lake to the left and rock face on the right. The average Italian male drives these narrow winding roads in his Alfa/Fiat at the fastest speed possible accompanied by the squeal of tyres and the sound of the horn, dismissing all in his path. That is until he sees, through his Ray Bans, the unmistakable profile of a Range Rover coming in the opposite direction, at full tilt, whereupon he dives for cover almost bowing in respect. At Belagio I indulged in an excellent five course meal and a comfy bed in the Hotel Perlo Panorama, overlooking Belagio, for an inclusive price of about £55 which seemed reasonable.

I had some work to do for tour operators, Page & Moy Holidays on my way south, so busied myself the next two days photographing some of the most staggeringly beautiful places I have ever been to, such as Verona, Siena, San Gimignano, Florence and Perugia. It was in the hilltop town of San Gimignano that the Range Rover had to show the other side of its character.

To get a shot of the magnificent mediaeval towers that dominate the skyline I had to do quite a bit of exploration to find just the right viewpoint. This turned out to be in the middle of a steep hillside allotment, inaccessible to any two wheel drive car.

The Range Rover easily transported me and all my equipment to exactly the spot I needed in air conditioned luxury, the temperature soaring already, even though it

All photography by the author on Fuji professional film.

The author working off the Range Rover's tailgate to gain height for a shot of a shepherd's house.

was only 7.30am. I took a quick Polaroid of the allotment tenant leaning on his hoe, which kept him happy, though the Range Rover, driven with care, treads very lightly on the soil, and left little evidence of its presence.

From here it was another blast south to my meeting with the University archaeologists, a jolly group numbering some 25 and most suffering from a hangover. They bore no resemblance to the storybook image of the archaeologist in his Harris Tweed jacket, half moon glasses, brogues and absent mindedness, but they all shared a passion for archaeology and relished digging in the hope of unearthing an important find.

We headed further up into the mountains to secret places known only to a few locals and Leicester University, where the group were excavating three sites of 12th and 13th century villages. The beauty of these places for the Land Rover enthusiast are that they are virtually inaccessible to all but 4WD vehicles and there is no-one to shout 'Oi! where do you think your going?'

Some of the old farm buildings are being converted into holiday homes for those desperate to escape from Rome at the weekends, and one such Roman, whose land a site was on, had a 1940's Willy's Jeep to play in, which he did with typical Latin style. I tried to explain in Italian and English that if I'd had my 80" Land Rover there we could have had a lot of fun, but 28 grand's worth of England's finest was not for playing with. He saw my point, had a look over the Range Rover and expressed great enthusiasm, accompanied by a lot of arm waving and gesturing.

My trips up to the hill top castles that guarded the area from attack back in the 12th century were the most sporting drives, along tracks more suited to a 13th century mule than its 20th century counterpart. Directed by archaeologist Paul Beavitt who was my client, we went places in the Range Rover he did not believe possible.

His cautious words 'you'll never get this up there' are, of course, the wrong things to say to an old Land Rover trialer. Oblivious to the value of the Range Rover, low ratio was engaged and in first or second we conquered all that the mountains could throw at us, accomplished in air conditioned splendour with cool drinks available from the Koolatron as and when needed.

On the last site we met up with Neil Christie, the co-director of the archaeological project, in the University's own 110 County Diesel Turbo and, in convoy, climbed up to the site of a 14th century farmhouse being excavated by a group of students. Much to their delight we revealed a secret supply of chilled Champagne in the Koolatron with my film. One by one they crawled out of the deep, dusty hole that had been their workplace for the day and savoured a glass or two, before piling into the Land Rover and

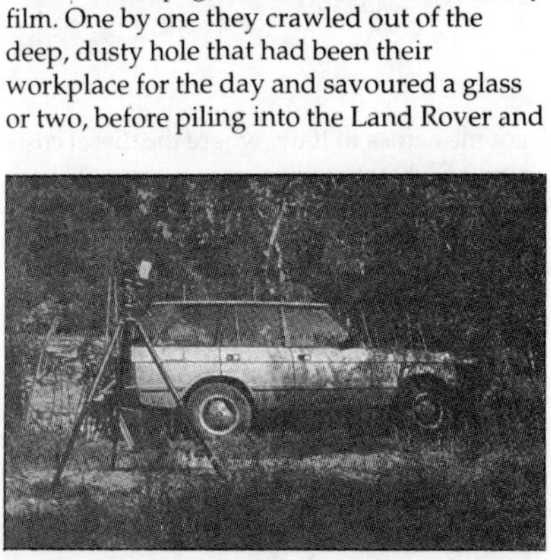

Working from an allotment to achieve the shot on page 150.

the Range Rover for the journey back to the school house which they were using as a base.

Because of the drought, there was no running water in this establishment so showering consisted of taking a washing up bowl full of cold water to the school shower room and, with a baby's potty dousing yourself in water, then a good soaping all over, before the final rinse with a further three potty's full. With practice it was very effective.

My work finished, it was time for the long trek home and, following the success of my outgoing trip, I decided on the same route back. The only highlight was on the German Autobahn. Shortly after sunrise, when proceeding at my usual sedate 80 mph I noticed a familiar profile bearing down on me at an alarming rate. Unmistakably a Range Rover, but as it closed in, an English registered one.

The 'K' reg Vogue S.E. shot past at a steady 100 mph, a family of four on their way home from holidays. Dad was at the wheel and gave me the sort of look that said 'come on then'. I took the bait, and floored the throttle cranking the poor diesel up to an indicated 100 mph, which was unfair after we'd been through so much together. It did it, albeit a little noisily and, for about five miles, the two of us drove in convoy at a steady 100-ish across Germany, before I chickened out and dropped back to my usual 80 to 85 mph, smug in the knowledge that he would be refuelling twice as many times as me on the return trip to England, and I'd probably be in Calais before him for that very reason.

I made Calais for a late afternoon sailing well before the one I was actually booked on. There was room aboard so they let me straight on, where it was a case of beating the backpack brigade to the comfy bench seats for an essential sleep to charge up my batteries for the final slog up to Leicester.

At the exit from Dover docks I picked up Steve and Sarah, a couple of friendly Glasgow hitchers, whose lively conversation about their holiday in Holland kept me awake until I dropped them off at the Leicester services on the M1 in the same drizzle I had left a week earlier. Just a few more miles and I was home to a hot bath and a large duty free brandy.

I had covered more than 1100 miles since leaving the archaeologists near Rome, 38

Δ
A view unchanged in seven centuries - the golden Rascino plateau in the Cicolano mountains.

Another of the many mule tracks easily ∇ negotiated by the Tdi Range Rover.

hours earlier, and it is testament to the comfort of the Range Rover Tdi that such a journey is possible, and all at a fuel consumption comparable with an average saloon car. No longer need the Range Rover be the preserve of the few who can afford the thirsty V8, here is a truly usable Range Rover to rival the appeal of the BMW's, Merc's and Jag's with the added bonus of being the world's best off roader.

• Koolatron 12 volt refrigerators are advertised in *LRO* and are available from: M.R.K. Ltd., Kingsgate House, 536 Kings Road, London SW10 0TE. Tel: 071 351 7444. Fax: 071 351 5014.

Crossing of Alps part of Hannibal's 1500-mile, five-month trek from Spain to Italy (see map, opp). He'd have been quicker in a Range Rover, which takes the winding gravel tracks in its stride

The Elephant Man

Phil Llewellin takes a Range Rover Vogue LSE across the Alps, tracing the footsteps of the elephant-toting Carthaginian war hero Hannibal

PHOTOGRAPHS BY IAN DAWSON

IAN BOTHAM AND BERNARD LEVIN have something in common, apart from membership of the human race. The beefy, beer-swigging cricketer and the apparently far from athletic intellectual, journalist, critic and opera buff have walked right across southern France and over the Alps to Italy. They followed in the bloody footsteps of Hannibal, the 29-year-old Carthaginian general who astonished and terrified imperial Rome by marching his army all the way from Spain and over mountains that formed seemingly impregnable defences. Hannibal's feat is made all the more extraordinary by the fact that his army included armoured fighting elephants.

Hannibal's exploits struck a resounding chord when Land Rover was wondering how to mark the 10th anniversary of Europe's biggest four-wheel-drive show. It attracts about 55,000 enthusiasts to Val d'Isère, 80 miles south-east of Geneva. The little town is 6000ft above sea level and spends most of the year catering for suicidal lunatics who hurtle down snow-covered slopes with planks strapped to their feet.

The week-long *Salon International 4x4 et Tout-Terrain* takes the town over. The main attraction from Joseph Publique's viewpoint is the opportunity to be terrified by professional off-roaders who tackle bowel-watering hazards on the spectacular Espace Killy proving ground, 4000 literally breathtaking feet higher than Val d'Isère. There are views of Mont Blanc's eternally snow-capped summit, 30 miles away, if you are brave enough to keep your eyes open.

My mission was to see how the Range Rover Vogue LSE – Solihull's answer to the stretch limo – fared on rough, steep alpine tracks. I had previously used one to explore the Canyonlands of Utah, where I came across a number of challenging obstacles, one of them a rock formation called the Lion's Back. What looks like a gigantic extrusion of petrified toothpaste, or an immense and steeply elevated howitzer, rises about 200ft in as many yards. The spine is very narrow and slightly curved, so there appears to be only fresh air between you and a premature appointment with St Peter. I renamed it Laxative Ridge. The long-wheelbase Range Rover crawled up and trickled down without missing a beat – more than can be said for the heart of its white-knuckled, sphincter-clenched driver.

Getting from Geneva to Val d'Isère by road takes the best part of four hours, because the few roads that wriggle through Europe's highest mountains are no straighter than a politician's answers. Land Rover saved time by organising an *Apocalypse Now* flight in six-seater Ecureuil helicopters. We flew below cloud-capped crags, past immense limestone ramparts streaked with waterfalls, over wooded whalebacks and, wonder of wonders, into a rainbow that formed a complete circle.

I thought about Hannibal and tried to imagine what his army would have looked like from the air. He was born in what is now Tunisia, right at the heart of the Carthaginian Empire, which stretched from present-day Lebanon and Israel to Morocco and Spain, where Hannibal spent most of his early life. He was still a boy when his father made him swear eternal hostility to Rome, Carthage's great rival.

The five-month, 1500-mile march to Italy started from the Spanish port of Cartagena. Nobody can be sure of the numbers involved, but Bernard Levin's book, *Hannibal's Footsteps,* talks of 60,000 infantry, 9000 horsemen and 37 elephants. Another source agrees about the pachyderms, but reduces the human element to 40,000.

None of my conversion tables equate horsepower with the earthquaking hefferlump variety that was Hannibal's answer to Rome's chariots. Photographer Ian Dawson and I had the 4.2-litre version of Land Rover's venerable V8 at our disposal, so the Vogue LSE should have provided 200bhp and 250lb ft of torque. I say 'should have' because high altitudes seriously sap the power of naturally aspirated piston engines. There were times when it felt more like 20bhp and 25lb ft.

A generation of drivers has matured since the first three-door, 3.5-litre Range Rover made its debut on 17 June 1970, priced at £1998. People old enough to remember the impact made by this pioneering vehicle's unprecedented blend of assets invariably lament the extent to which the Range Rover's price has increased, even when more than two decades of inflation are taken into account. The Vogue LSE is now listed at 19.21 times more than the original version cost in 1970. The price of today's cheapest Rolls-Royce is only 10.59 times higher than it was then. The figure for the basic Mini is 7.32.

That said, the current Range Rovers are light years removed from their ancestor. The design philosophy has remained constant, but specifications have changed to embrace such features as bigger engines, automatic and five-speed manual transmissions, traction control, air-conditioning, anti-lock brakes, multi-mode air suspension, five doors, leather upholstery, discreet applications of polished wood and, of course, the long-wheelbase model.

Increasing the luxury, the price and the cachet, year after year, has been the marketing strategists' profitable reaction to the Range Rover as a peerless status symbol. The fact that very few owners ever tackle anything more demanding than the stately home's pristine gravel drive epitomises the difference between need and want, image and reality. The typical owner is like the rich person who buys a very expensive camera to complement his lifestyle, but eschews its many functions in favour of the automatic setting.

The LSE's wheelbase is eight inches longer than the SE's, so my tall, lanky colleague could pray in comfort in the back seat while I tried not to slither over precipices. There's enough room for a Harlem Globetrotter to stretch his legs while wearing stilts, but a lot of the acreage is wasted. A bigger boot would be far more useful, because the Range Rover's battleship bulk belies a relatively small capacity for concealed luggage.

My experience of Range Rovers goes back about 20 years, but it always takes me a little time to adjust to the vehicle's steering and cornering characteristics: they can be quite alarming, particularly if you

Scenery dramatic. Careful use of ABS partly alleviates running-away sensation felt when descending steep slopes in auto off-roaders. Llewellin (above) found headroom marginal

were expecting this beam-axled showboat to handle like a conventional car. That said, the gap has narrowed since height-adjustable air suspension replaced coil springs in 1992. It should have happened a year or two earlier, but the development engineers were not satisfied about reliability and social acceptability. The genteel sighs that now emanate from the airbag suspension when you come to a halt are nothing to the tremendous, avalanche-triggering farts it emitted in the early days.

There's a strong hint of Citroën DS19 about the way the LSE relaxes at a standstill. This is part of a five-position system that spans almost six inches. The access mode makes getting in and out that bit easier; low-profile is automatically adopted when speed tops 50mph; standard speaks for itself; a button selects high-profile for off-road driving; extended profile can lower a wheel far enough to obtain traction in situations where a grounded Range Rover might otherwise be stuck.

Taking it easy was the name of the first game. The narrow road from Val d'Isère to the Col de l'Iseran looks like the result of an explosion in a spaghetti factory. Signs warning motorists to beware of low-flying angels would be appropriate as you climb to 9085ft. A memorial at the top of the pass tells how the road took 20 years to build and was completed in 1937. It was the highest in Europe until 1962, when the road over the Col de la Bonette between Nice and Briançon topped it by just 108ft.

How well a vehicle performs off-road depends to a great extent on individual perceptions, experiences and expectations. Dawson has spent very little time motoring up and down steep slopes surfaced with mud, gravel, grass, rocks, running water and the bleached bones of dead sheep. I could understand why he was making amazed, appreciative and slightly apprehensive noises in conditions that the Vogue LSE was handling without unleashing more than about 25 percent of its potential.

What impressed me was not so much what it did as how it did it. Any off-road vehicle could have completed the route, which took us through sensational scenery to a vantage point on the Italian border at Mont Cenis. Indeed, we shared one track's hairpins with nothing more specialised than a Fiat Tipo, a Volvo 460 and a Citroën AX whose driver, complete with beret and cigarette, looked old enough to have witnessed Hannibal's march through the mountains. The track was firm – it probably dated from Napoleonic times and was built for military reasons – but was definitely rough and steep enough to daunt the average Brit in anything without four-wheel drive and good ground clearance.

Experience tells me that the Vogue LSE keeps going in conditions that would stop most of its rivals. Cross-axle differential locks could make it even better, in my opinion, but would be wasted on 99.999999 percent of owners. What makes the stretched Range Rover so exceptional is the way its long wheelbase and air suspension cosset driver and passengers. The chassis's ability to absorb punishment is difficult to quantify, in human terms, but can make a significant difference to your physical state at the end of a long, low-gear day. Fingers, wrists, arms and shoulders are not the only parts that benefit, for you find that your stomach muscles spend less time tensed to combat the sharp vertical and lateral movements inflicted by most other vehicles.

Talking of which, it was good to encounter a stark, hard-riding, 1945 Willys Jeep – a fine example of the Range Rover's spiritual ancestors – after we had reached the huge and desolate Fort de la Turra at almost 10,000ft above sea level. Lower down the track, a stone dated 1891 revealed the Route de la Turra that was to have been built by mountain troops known as *Chaseurs Alpins.* A nearby monument commemorated comrades who had been killed in an avalanche.

The Col du Mont Cenis is one of several contenders for the place where Hannibal and his Carthaginians crossed the last major barrier before marching into northern Italy and storming what is now Turin. Snow, landslides and hostile natives had reduced the army to about 20,000 infantry, 6000 cavalry and only a few elephants by the time they reached the Po valley.

Hannibal was destined to spend the next 15 years in Italy. He won two classic battles – Cannae and Trasimene – campaigned as far afield as Taranto and marched to within sight of Rome before returning to Carthage. He was about 64 and living in exile when he took poison rather than be taken prisoner. His last words, as quoted in *Hannibal's Footsteps* were: 'Let us now put an end to

the great anxiety of the Romans, who have thought it too lengthy and too onerous a task to wait for the death of an old man whom they hate.'

A modern historian, Gavin de Beer, postulates a fascinating parallel in his book, *Alps and Elephants*. He likens the Italian campaign to Hitler's most charismatic commander, General Erwin Rommel, landing 20,000 troops in Scotland in 1940, marauding as far south as Plymouth and beating several British armies before sailing home undefeated in 1955, to help defend the Fatherland.

Nothing defeated the Range Rover on this two-day challenge, but one of its most important features did let me down. It was mid afternoon. Laser-bright light was slicing through the thin, cool air as we climbed a narrow, loose-surfaced track whose eternity of steep, tight hairpins could have been slashed into the mountain by a rapier.

Its considerable presence makes it easy to misjudge the long-wheelbase Range Rover's size. It is actually only an inch longer than a Ford Mondeo estate, and four inches narrower. Be that as it may, there was too much vehicle and too little terra firma for several of the hairpins to be negotiated in fewer than three bites. Getting it wrong while reversing would have been the last of life's mistakes. I failed to take a wide enough swing at one of the corners, slithered back a foot or two, had another go and slid back again before taking a better line. The automatic traction-control system had failed to report for duty on the one occasion when it was really needed. We were in no real danger, but equipment failures can be fatal.

At the other end of the scale, the first anti-lock braking system devised for a vehicle of this type proved its worth more than once. Land Rover's experts have always taught off-roaders to tackle downhill slopes by selecting a low gear and relying on the engine for retardation. Dabbing the brakes can lock a wheel, precipitating a slide that slithers the vehicle sideways and, if the slope's even moderately steep, may cause it to capsize. Sensible use of the ABS system virtually eliminates this risk, and also helps compensate for the slight 'running away' sensation I always experience when driving down steep slopes in off-roaders with automatic transmission.

This is one of the tallest non-commercial vehicles on or off the road, but the sunroof's housing made headroom marginal for your short-legged, long-bodied reporter. And while we're on the subject of the interior, the ergonomics are Stone Age – a reminder of the basic design's antiquity. Part of the dash's plastic trim appeared to have been gnawed by rats, and one side of the instrument binnacle was half-an-inch out of line. I would have gone straight back to the dealer had I just signed a cheque for the thick end of £40,000.

Back at Val d'Isère, we exchanged tales of high adventure as we drank beer in the Hotel Altitude. I concentrated on reading the ski resort's tourist propaganda and was puzzled by a reference to piste grooming machines. According to a colleague, they are Heath Robinson contraptions that make you look neat and tidy when drunk. He also reckoned that Land Rovers have winches in order to pull muttering rotters out of pubs. Hannibal would have used them to haul elephants over Alps.

Useful kit on lwb Range Rover includes traction control. At altitudes of 10,000ft, its 200bhp is diminished. Land Rover organised event to mark 10th year of Europe's biggest 4wd show

ROYAL COMMAND PERFORMANCE

Shogun's 205bhp V6 is spirited but lacks V8's torque

Range Rover puts 4.2-litre V8's muscle to good use

Country folk like the Berkeleys are used to the unpredictable upsets that go along with dogs 'n' guns and fields of green. Like the time King Edward II got himself murdered most foully in one of the guest rooms. Or the time Oliver Cromwell came crashing through the front wall with his mates, spoiling for a fight.

You see, the Berkeleys have been upholding the musty, dog-haired traditions of country life for more than 850 years. And their home is their castle. Literally.

Huge 17th century wood and iron doors draw back to welcome us into the courtyard. Nosing its way in, ever so self-consciously, is the grandest, most expensive, most powerful Shogun yet to turn a wheel in this country. At £35,899, this flagship Mitsubishi costs £9500 more than its weaker three-litre siblings, and delivers a stunning 58bhp more than the 'small' engined Shogun with its 3000 GT-based 205bhp 3.5-litre V6. Sucking in the courtyard air once enjoyed by kings, the Shogun makes no apologies, acknowledges no peers. It is, simply put, the most powerful off-roader on sale in the UK. It is damn near the most expensive, too — bar one.

When the doors finally creak shut, there is another massive presence in the courtyard. No self-consciousness among the cobblestones here, this one feels completely at home parked nose-to-nose with eight centuries' worth of countrified money and privilege. It is, appropriately, the second most powerful off-roader in the UK; and, at £39,160, the Range Rover LSE is the most expensive.

Mitsubishi admits that it is baying for Range Rover blood with the latest Shogun and that on power, specification and refinement, it deserves to tilt against the best there is.

PHOTOGRAPHS BY STAN PAPIOR. BERKELEY CASTLE OPENS TO THE PUBLIC FROM APRIL TO OCTOBER

Relatively car-like handling helps Shogun's cause

Range Rover rolls dramatically and ride is often crude

As befits its aspirations, the 3.5-litre Shogun's equipment list includes all of the goodies found on the top-spec three-litre model plus a few potent extras like anti-lock brakes, air conditioning, CD autochanger and electrically adjustable front seats. It also shares new alloy wheels with the rest of the range and, unusually for an off-roader, a driver's airbag graces the steering wheel. But the power is the thing here.

Just take a look at these figures. This V6 thumper produces 205bhp at 5000rpm and a root-pulling 221lb ft of torque at a relatively lazy 3000rpm compared with the three-litre Shogun's 147bhp and 174lb ft. The result is quickness that belies the Shogun's hugeness; the 0-62mph sprint is claimed to take just 12.5sec — and we think that's unduly pessimistic.

The hairiest Shogun will need every ounce of that power to go Range Rover hunting, though. The LSE's 4.2-litre 200bhp V8 just about equals the Shogun's power output but, crucially, pumps out a huge 250lb ft of torque at 3250rpm (even plain old Range Rover Vogues manage 231lb ft). At the sharp end, that means the biggest Range Rover will bellow its way to 60mph in a swift 10.8sec. Given the mammoth frontal area with which the Range Rover and Shogun cleave the air, their top speeds of 110 and 113mph respectively are damned impressive, too.

The Shogun can be ordered with a manual five-speed 'box or, as tested here, a multi-mode four-speed automatic. In 'power' mode, the revcounter climbs higher in each gear range and the 'box kicks down with less right-foot flexing, while in 'economy' mode the Shogun just lazes around. For ultra slippy getaways, punch the 'hold' button and the Shogun will pick its way gently from standstill in second gear.

The Range Rover, on the other hand, does quite nicely without 'modes' and a multiplicity of buttons, content to let its massive torque carry the day.

Externally, latest Shogun is bigger in every direction than Range Rover — a lot of metal for £35,899

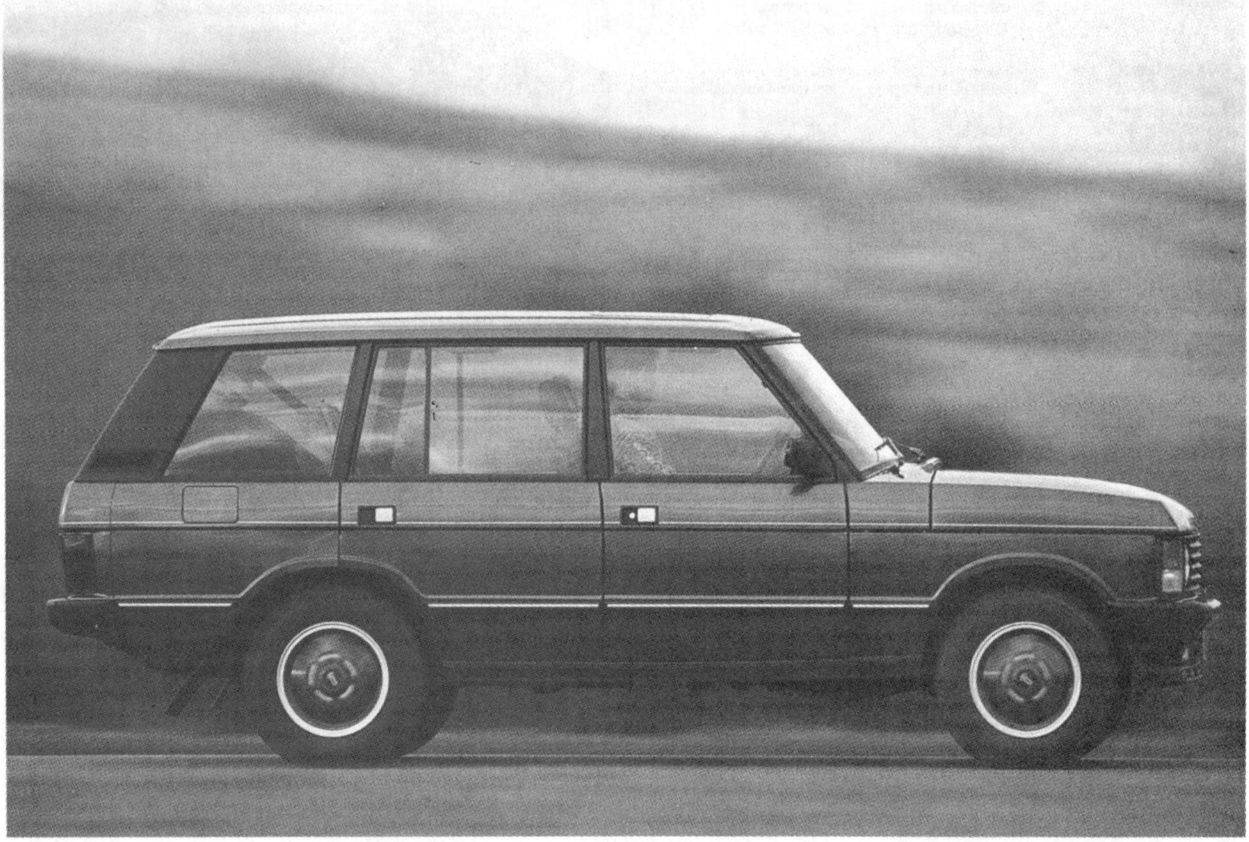

LSE is eight inches longer than standard Range Rover. It's the most imposing, outrageous Range Rover yet

Factfile

	Mitsubishi Shogun 3.5	Range Rover LSE 4.2
How fast?		
0-60mph	12.5sec *	10.8sec
Top speed	113mph *	110mph
Overall mpg	18.5	16.7
* Manufacturer's claimed figures		
How much?	£35,899	£39,160
On sale in UK	now	now
How big?		
Length	4725mm (186.0in)	4648mm (183.0in)
Width	1785mm (70.2in)	1778mm (70.0in)
Height	1895mm (74.6in)	1835mm (72.0in)
Wheelbase	2725mm (107.2in)	2743mm (108.0in)
Weight	1985kg (4367lb)	2150kg (4730lb)
Fuel tank	92 litres (20.4 gall)	82 litres (18 gall)
Engine		
Max power	205bhp/5000rpm	200bhp/4850rpm
Max torque	221lb ft/3000rpm	250lb ft/3250rpm
Specific output	59bhp/litre	47bhp/litre
Power to weight	103bhp/tonne	93bhp/tonne
Installation	longitudinal, front, four-wheel drive	longitudinal, front, four-wheel drive
Capacity	3497cc, 6 cyl in vee	4278cc, 8 cyl in vee
Made of	aluminium alloy head and block	aluminium alloy head and block
Bore/stroke	93mm/86mm	94mm/77mm
Valves	4 per cyl, dohc	2 per cyl, ohv
Ignition and fuel	multi-point injection and electronic ignition	Lucas L-Jetronic injection and electronic ignition
Gearbox		
Type	four-speed automatic	four-speed automatic
Ratios/mph per 1000rpm		
1st	2.80/6.5	2.48/7.9
2nd	1.53/11.9	1.48/13.2
3rd	1.00/18.3	1.00/19.8
4th	0.75/24.3	0.73/26.9
Final drive ratio	4.64:1	3.54:1
Suspension		
Front	independent double wishbones, torsion bar, anti-roll bar	live axle, air springs, anti-roll bar
Rear	multi-link solid axle, coil springs, anti-roll bar	live axle, air springs, anti-roll bar
Steering		
Type	recirculating ball, power assisted	worm and roller, power assisted
Lock to lock	3.3 turns	3.8 turns
Brakes		
Front	276mm ventilated discs	298mm ventilated discs
Rear	315mm discs	290mm discs
Anti-lock	standard	standard
Wheels and tyres		
Size	7x15in	7x16in
Made of	cast alloy	cast alloy
Tyres	265/70 HR15	205 R16 Michelin XM+S
Sold by	The Colt Car Company, Watermoor, Cirencester, Glos GL7 1LF. Tel: 0285 655777	Land Rover Ltd, Lode Lane, Solihull, West Midlands B92 8NW. Tel: 021 722 2424

Choice of three suspension settings for Shogun drivers

Facia has too many gimmicks and not enough style

Fabulous equipment level includes driver's airbag

Rearmost seats fold away to provide versatile load area

Air-sprung Range Rover is better in a straight line

Facia is angular, old-fashioned, but not without charm

Leather and wood cabin feels majestic and special

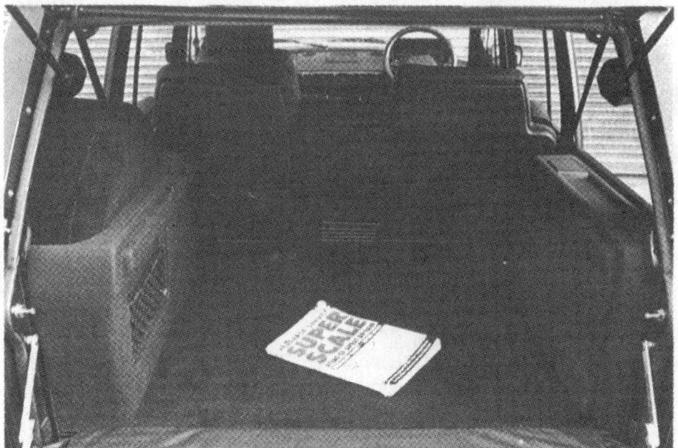

Extra length improves rear legroom but not load space

The Shogun delivers as much passing, merging and high speed cruising ability as anyone driving portable real estate has a right to expect. At a comfortable cruise, the Shogun is an especially pleasant companion, loping around the 20mph increments with power to spare. Push hard, though, and the V6 begins to sound ragged and you won't ever find yourself revving for the hell of it. The Shogun will find the LSE's smooth, refined V8 a tough act to follow; if its power delivery were breakfast it would be Shredded Wheat; the Range Rover's, on the other hand, is a steaming plateful of tapioca pudding.

Push the LSE around a bit and you'll pay for it, though; we recorded a fairly dismal average of less than 17mpg. The Shogun proved the more frugal mount, bettering 18mpg.

Whereas the Range Rover subscribes to the roll-and-lurch school of dynamics, the big Mitsubishi is much more car-like in its handling. A high centre of gravity militates against flat cornering, but the Shogun won't throw any nasty surprises at you, understeering on its massive 265/70 tyres with lots of early warning. To cover all the bases, the top dog Shogun's suspension has three settings of soft, medium and hard. 'Soft' delivers a wallowy, sloppy, albeit absorbent ride and 'hard' delivers much less roll, but a rather unpleasant, sudden impact over the rough stuff. The 'medium' setting is a pretty good compromise.

Approach either of these cars from any angle and you are struck by, well, by their outrageous hugeness. Big? No, titanic is more like it. The Shogun actually sits 30mm (1.2in) higher than its smaller-engined siblings, rudely displaying wheel wells that could swallow a GTi whole. Mitsubishi explains that rather than lose ground clearance after shoe-horning the taller 3.5-litre V6 into place, it simply increased ride height to maintain the Shogun's imperial altitude over the mud.

Once you have managed to scale the door sills and climb into the seat, an ego expanding sense of grandness will creep over you. You will lock eyes with artic drivers without flinching. To increase that sense of mastery, an altimeter, inclinometer and compass will blink and wobble at you from the facia.

And while the Range Rover is plenty big, the Shogun is bigger. It is 30mm taller than the LSE, it is 10mm wider and it is, if you count the huge spare wheel carrier, 77mm longer. Press the button, though, and the LSE will kneel an obsequious 30mm for passengers while the Shogun stands proud, oblivious to splayed legs and broken heels.

Let's face it, both the Shogun and Range Rover are about 10 times better as off-roaders than they will ever need to be. Even Mitsubishi admits that someone spending £36K on their wheels is pretty unlikely to risk bending it in farmer Brown's back forty. But if you must have proof of the Shogun's considerable off-road ability, simply refer your bar stool colleagues to the numerous Paris-Dakar rally class wins Mitsubishi has totted up since 1983.

"It just doesn't have the majesty of the Range Rover," said one tester from the back seat of the Shogun. It isn't just the tweedy familiarity, either; the wood and leather trimmed cabin, sharp angled and not the least concerned with passing fashion, feels traditional, stoically upright, conservative even. Spend time in the Shogun and you begin to resent those arcade-game altimeters and inclinometers gimballing about uselessly on the dash. Complain about the execution all you like, though, because there is no denying that, for the money, the Shogun's cabin lacks for virtually nothing. And the five-door boasts what must be the biggest glass sunroof in the business; the thing feels almost like a convertible.

Against the Range Rover's ageing, angular suit of armour, the Shogun offers a sleeker, better value, lower depreciating package for the unsentimental, streetwise buyer. There are, however, the intangibles to consider.

The Berkeleys don't look much like car enthusiasts, but the bluff profile of the big Range Rover is something they instantly recognise and understand. Mrs Berkeley is politely interested in the Shogun, asking where it is from and is it a good thing to have.

For what it's worth, people with castles seem to prefer Range Rovers. ■

SIZE AND POWER

Dave Barker catches up with West Yorkshire's Motorway Patrol Range Rovers

ALMOST SINCE its launch, police forces throughout the country have used the Range Rover in a number of roles. Most commonly, because of the Range Rover's size and power, as a motorway patrol vehicle.

After reading James Taylor's 'Rare Rover Tour of Duty' (last issue March 94) featuring Solihull's police demonstrator of the mid 1970's, I went along to meet the West Yorkshire police motorway section and to see today's police Range Rovers, 23 years on. What changes have been made over the years and how is the Range Rover coping today with the changes and increased speeds and traffic on our motorway networks?

On a very foggy and icy morning, with a good covering of snow still on the ground, I arrived at the West Yorkshire police motorway sections headquarters, sited just off the M1 near Wakefield, to meet Inspector Alan North who had kindly offered to show me his Range Rovers.

Because of the fog and ice there was a feeling around the motorway section that, because of the day's very bad driving conditions, another big motorway pile-up wouldn't happen again today. Inspector North explained on a large scale map the area of the M1, M62 and A1 that his men and vehicles patrol includes some of the most exposed stretches of motorway in the country. This includes the M62 out of West Yorkshire over the Pennine's to the Lancashire border at a place aptly named Windy Hill (a section of motorway that on a couple of occasions each year is passable only in a 4x4).

In total the West Yorkshire motorway section covers some 240 carriageway way miles. The motorway sections have their base just off the M1, close to junction 41 just south of the busy junction of the M1 and the M62 at the centre of the operating area, and at the very heart of the motorway network in West Yorkshire.

At present the motorway section operates three Range Rovers out of a total fleet of 30 vehicles, the rest of the vehicles being a mixture of cars, Rovers, Vauxhalls and Fords (including some high performance Ford Cosworths). They also have three brand new Range Rovers, at present in the force's own workshops being kitted out to their own specifications as new additions to the fleet. By the time you read this article, West Yorkshire will have all six operational.

Inspector North explained that he regarded the Range Rovers as the workhorses of the fleet, each vehicle will spend approximately 12 hours a day out on the road and cover some 350 miles each day. It's the increased carrying capacity that the Range Rover offers over normal saloon cars that makes it such an ideal vehicle as a motorway patrol car.

The Range Rover will carry, in addition to the standard equipment all the motorway cars carry, many additional items such as the larger type road cones, warning boards and other equipment that may be needed at the scene of an incident or a motorway accident. Because of its physical size and power, the Range Rover makes a far more versatile vehicle at the scene of a motorway accident; with the addition of a high level telescopic lighting mast, the Range Rover is far more visible from a distance.

With motorway speeds today being so much faster, and the roads carrying more traffic, the more visible the police vehicle is the quicker other drivers will slow down and so the safer the officers are. Just a few added seconds can make ali the difference to a vehicle slowing down and missing something stuck in the middle of the motorway or driving straight into it.

With its power, the Range Rover is also able to clear the motorway of any wreckage, to push a damaged vehicle onto the hard shoulder or even tow a broken down lorry out of the way. The aim is to get the carriageway open again as fast as possible, hopefully preventing any additional accidents further down the road as faster vehicles start to run into the standing traffic.

One other 4x4 vehicle manufacturer that was offering vehicles to the police to evaluate, withdrew their offer when they were asked to fit towing points just in case the vehicles were needed to tow a lorry off the carriageway onto the hard shoulder!

When new Range Rovers are delivered to the West Yorkshire police, apart from factory fitted, up-rated alternators and springs and with some of the luxury items such as carpets missing, they are standard vehicles. They then go into the force's own workshop/garage to have all the necessary police items fitted. Over the years, the workshops and mechanics have become experts in fitting and adapting the more specialised equipment needed in today's motorway police vehicles, which now include such things as video cameras and satellite tracking systems.

Looking at the photographs from the 70's in James Taylor's article, some of the police fittings seem to have changed very little over the years. The raised telescopic mast seen on the Range Rover demonstrator in the 70s is still is service today but is being replaced as budgets allow newer equipment to be purchased. The older style mast still works well but is felt to be a little too flimsy for safe use on some of the more higher and exposed windswept stretches of the M62 in West Yorkshire, as the mast is only fitted to the roof of the Range Rover. The newer type hydraulic mast now being preferred is mounted through the roof of the vehicle, right down to the chassis, so making it far more stable in high cross winds.

A new addition on the dash is the 'Traker' system which enables a police vehicle to track and position via satellite a stolen car which has been fitted with a Traker transmitter with a very high degree of accuracy. The system is still in its early days, and still being tested by many police forces.

Twenty Three years on, the Range Rover is still regarded by the many polices forces as one of the best vehicles on the motorways. But don't forget, they also operate a number of un-marked vehicles as well!

My thanks to Inspector Alan North and to Chief Inspector Melvyn Bunting of the West Yorkshire Police for all their time and help.

Telescopic mast houses blue flashing beacon and floodlighting ▷

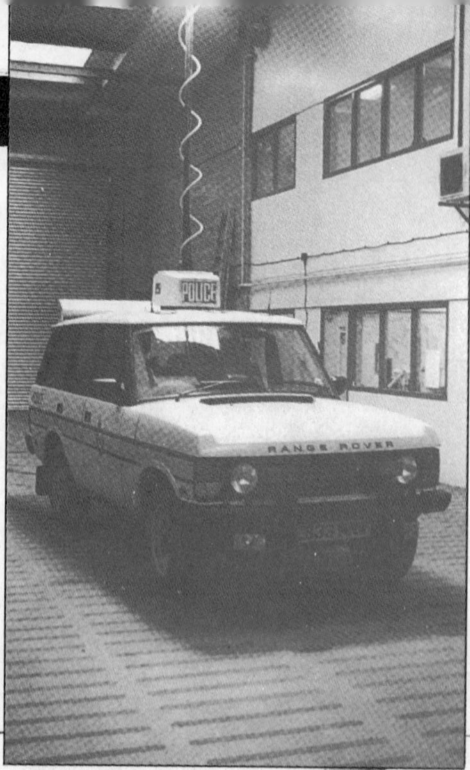

▽ ***Interior includes Traker - a satellite system used to trace stolen vehicles***

▽ ***Highly visible from the rear***

RANGE ROVER

The Rover Boys and their Electronic Suspension

BY RAY THURSBY

PHOTOS BY RON SESSIONS

Only the British could have created the Range Rover. Where else but in a land where dignity and ability coexist with—and are sometimes the same as—silliness could a vehicle like this one be developed? Range Rovers combine the attributes of Winston Churchill and Benny Hill and, what's more, blend them in perfect harmony.

On one level, you might question the Range Rover's suitability for all-terrain travel. Of what use to an off-roader are

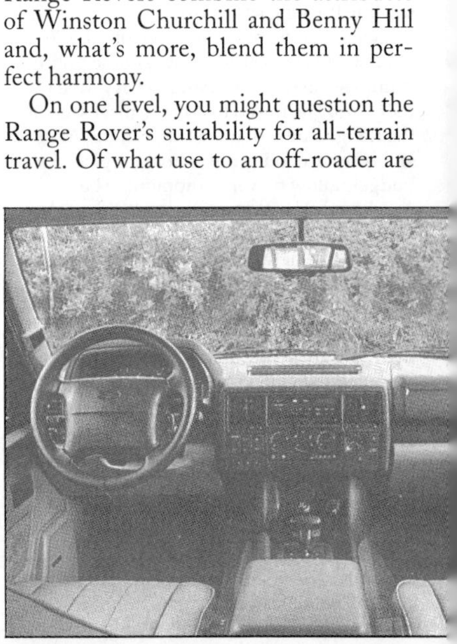

■ Amid the splendor of leather and burled walnut, '95 models sport dual airbags, dual-zone climate controls and dual cupholders in an all-new instrument panel.

leather seats, plush carpeting and shiny wood trim? What does a lustrous paint job lovingly applied over aluminum coachwork contribute to utility? The answer is *not much*; my inclination when first taking one into the desert was to find something to hit, just to get the initial dismay one feels after damaging such a jewel over and done with. But that was merely a case of temporary insanity, and the maneuverable Range Rover went about its business without acquiring any rock rubs or cactus burns at all.

Many buyers never will drive their Range Rovers on any surface more demanding than Rodeo Drive. For them, comfort and image are paramount, and these they get in quantity. Given the chance to ride high above the crowd on hide-trimmed thrones, driver coddled by power-assisted everything, passengers soothed by lots of sound insulation, automatic climate control and compliant suspension, who would not be charmed?

Around town, the Range Rover allows—no, encourages—one to revel in the kind of British luxury generally found in that other R-R. There is, simply put, no downside to driving a Range Rover through the wilds of Beverly Hills. Or being seen in one. Pretentious? *Moi*?

Nor is there a downside to driving one across the Gobi Desert or through the Superstition Mountains. Under the silk is strength built around a boxed steel ladder chassis and welded underbody unit. The hardware exudes an air of indestructibility that few other vehicles—including some previous Range Rovers—can match.

Choosing the Range Rover that's right for you couldn't be easier. If, that is, you fancy a County LWB (Long Wheel Base); last year's shorty is gone. That's not the only news to report, as the County's been given a new dashboard for 1995 (yes, it's already 1995 in Range Rover-land).

The resulting changes aren't merely cosmetic, though there is more burled walnut trim in the cabin this year. The new dash holds a passenger's airbag (with another in the steering-wheel hub for the driver) and a climate-control panel that's less complex than before, despite the addition of separate temperature dials for driver and passenger.

Although slightly heavier (by 227 lb) than last year's version, the County moves along right smartly, thanks to its 200-bhp V-8 engine. Originally produced by General Motors back in the 1960s, the all-alloy powerplant has been enlarged to 4.3 liters, equipped with the latest in electronic management systems, and feels as modern as any engine you'll find described in this magazine. Its partner is a smooth-shifting 4-speed automatic transmission connected to a full-time 4wd—via center differential and viscous coupling—unit with 2-speed transfer case.

Another noteworthy Range Rover feature is its electronically controlled air suspension. Pushbuttons allow the driver to choose among Access (lowers the vehicle for easier ingress and egress), Low Profile (high-speed driving), Standard, High Profile (off-road), and Extended Profile (to help you tiptoe away should you get high-centered on a rock). The air springs also give the Range Rover an extraordinarily good ride under any condition.

Other standard chassis features include power steering, 4-wheel disc brakes with ABS and electronic traction control. On the luxury front, every Range Rover is fitted with (in addition to power everything and automatic climate control) cruise control, a power sunroof, 8-way electrically adjustable heated seats with position memory, a new-for-1995 160-watt audio system with cassette and 6-disc CD changer, a central locking/keyless entry system, heated windshield and rear window, a trailer hitch and lots more.

Given the enormity of the equipment list and the quality of what's underneath, Range Rover's pricing seems almost reasonable. If you remain unconvinced, a test drive just might change your mind. You may feel as though you're riding in a fishbowl (the list of rich and famous who own these machines ensures that High Profile is more than just a suspension setting!), but you'll quickly discover that the dual-personality Range Rover may be the best all-around vehicle you can buy.

SPECIFICATIONS

Base price, base model	$52,500
Country of origin/assembly	United Kingdom
Body/seats	4D/5
Layout	F/4wd
Wheelbase	108.0 in.
Track, f/r	58.5/58.5 in.
Length	183.0 in.
Width	71.4 in.
Height	70.8 in.
Minimum ground clearance	8.0 in.
Curb weight	4807 lb
Maximum towing capacity	7700 lb
Gross vehicle weight rating	6019 lb
Cargo capacity	36.2-82.8 cu ft
Fuel capacity	23.4 gal.
Fuel economy (EPA city/highway)	12/15 mpg
Base engine	4.3L 182-bhp ohv V-8
Bore x stroke	94.0 x 77.0 mm
Displacement	4278 cc
Compression ratio	9.0:1
Horsepower, SAE net	200 bhp @ 4750 rpm
Torque	251 lb-ft @ 3250 rpm
Transmission	4A
Final-drive ratio	4.27:1
Suspension, f/r	live/live
Brakes, f/r	disc/disc, ABS
Base tires	205R-16
Steering type	worm & roller (p)
Turning circle	39.4 ft
Warranty, years/miles:	
Bumper-to-bumper	3/42,000
Powertrain	3/42,000
Rust-through	6/unlimited

Photography by Michael Bailie

Strange Rover

The quest for greater on-road sophistication and speed in off-road vehicles has led to this, the Overfinch Range Rover 570Ci. The ultimate road-burning 4wd? Mark Hales finds out

It's nearly a quarter of a century since the Range Rover first offered farmers the potential for visiting market in the same vehicle they used to cross their fields. The first ones had washable plastic seats and rubber floor covering so that the hosepipe could remove those parts of the farm brought in by wellingtons, and there was a farmer-familiar Land Rover transmission that clonked and whinnied most of the time. Two and a half decades on, the Mk4 Zodiac and Austin Cambridge have long since shuffled off to a place they deserve, but the Range Rover's exterior shape is still infinitely recognisable. Not only that, it shows no sign of impending decline.

You might however be tempted to ask why. Fact is that 99 per cent of Range Rovers never venture off-road, not intentionally at least. They're much more likely to be spotted blocking the road outside the private school at a quarter to nine, or speeding the local estate agent and his waxed jacket to the next business lunch. No likelihood of any mud on the upmarket pile carpet and velour interior there. So if you don't use the thing in conditions where it's likely to shine, what sort of a road car have you got?

Let's take a critical look at the 1994 model's specification, starting with the engine. The original was designed in 1952, with eight cylinders and one camshaft, and the latest pushes out 181bhp from 3.9 litres. At each end of the body, a huge dumb-bell between the wheels means the suspension is non-independent. Cabin room is available for five as long as they're not too big and a top speed of nearly 100mph is possible in return for one gallon of unleaded every 15 miles or so. All carrying a price tag for a well equipped auto version of £36,130. Unimpressive, huh? Makes a BMW 525iSE look like a bargain at £25,850.

Apart from fuel injection and an auto transmission option which you can see, and a change to the transfer gearbox which you can't, there hasn't been much in the way of radical change since 1971. So, either 4845 buyers in 1993 didn't care about technology, performance and the nation's fuel bill or the Range Rover is concealing other strengths under its boxy superstructure.

Then again, perhaps not. The notion of a true dual purpose on- and off-roader could never really be a feasible one, because if you really do want to plug the mud, you need to change tyres and then you can't travel fast on the road anymore. Back on its road rubber, the Range Rover is simply an overly large five-seater with moderate performance and high fuel consumption. All this, though, appears to be irrelevant and maybe the Range Rover's failure to evolve is an ironic constituent of its success. For different reasons, it looks as different now as it did in 1971 and the metamorphosis from trend-setter to anachronism has thus been achieved by accident and without any restyling. Over the years the Rover's two-box shape has gradually become a badge of tweedy excellence, an institution like that other great evergreen, the Mini.

But even someone with the most dedicated cynicism towards the breed — like me — would have to admit there's more to it than that. There *is* an undeniable something about the view from the driving seat. Not an immediately overwhelming thing — you need to get used to the swaying masses of that boxy body first — but a gradual feeling of exalted superiority that comes with a view across the average executive saloon's roofline. He can travel slightly faster, but he can't see what I can. This, and all that reassuringly solid metalwork around you, slowly seduces. Like a tank with windows, but without its tracks. A Centurion couldn't of course travel fast on the road with tracks.

If we accept the on-road penalties incurred by the Rover's off-road capability, there is still plenty of room for improvement in all aspects of its tarmac performance. The manufacturer offers a stretched 4.2-litre Rover V8 to power the stretched body of the LSE and for just under £40,000 this promises 200bhp and a top speed of around

● ***Modified steering is quicker and has sharper responses. Hides the OverRover's bulk***

110mph; the engine option is not available in the shorter wheelbase versions though. So if you must, turning your Range Rover into a real roadburner will need specialist attention, and if you want it done properly, a fair bit of cash. And there's nowhere better to spend your pile than at Overfinch, the Hampshire-based concern which has done more performance work to more Range Rovers than anybody else in the last 10 years.

The company will attend to absolutely anything on your RR but its reputation has been made primarily as the 'inventor' of the Chevrolet engine swap and it has converted an average of 30 Rovers every year since 1982. Out goes the Rover's aluminium 3.9-litre engine and in its place Overfinch has installed an iron small-block Chevrolet V8. It's a simple development of the no-substitute- for-cubic-centimetres theme, drawing simple and cheap hardware from America, the country which invented the big V8. The 570Ci test car featured 5.7-litres of understressed low-revving American ironware developing 285bhp, and a simply massive 344lb ft torque peak. If a 100bhp advantage over the Rover 3.9's 181bhp is not enough, there's a 5.7-litre Si version with aluminium heads and high compression pistons which puts out more than 300bhp.

The Overfinch's underbonnet view is neat and simple, and the Chevrolet engine looks as if the manufacturer had installed it, an observation that Overfinch regards as the best possible compliment. A free-flow stainless exhaust is the tailpiece to the engine conversion but as yet there's no catalyst as standard; the engine runs exclusively on unleaded fuel, however, and the company says that it is possible to fit one.

The GM V8s used in the conversion are assembled from a selection of Chevrolet performance parts rather than any aftermarket items and the engines are all built at Overfinch's Alton base. Drive is taken through the standard ZF four-speed auto box which ZF and Overfinch have adapted to suit the Chevy's rev range, and the basic engine conversion as we tested it would cost you £11,830 — assuming the salvage of usable bits from your existing car. This includes the exhaust and a basic uprating of suspension components plus a set of 255/60 Avon road tyres, some interior enhancements and all installation. Reasonably enough, it doesn't cover replacement of worn-out bits on your second-hand example. On top of that, the test car featured further tweaks to steering and suspension, leather-trimmed Recaro bucket seats with electric adjustment, and yet more styling additions inside and out. This adds another £6276 to the price of a new Vogue SE auto, for a total of £54,236.

This is a lot of money. Can it be worth it? If you must have a Range Rover and want it to go faster in safety, then yes it is. We managed 60mph from rest in 7.5 seconds which is a massive three seconds quicker than the standard 3.9, and a touch faster than our executive can expect from his 525 BMW. The magic and highly illegal 100mph flashed up in a whisker over 25 seconds which sees the OverRover lagging a little behind the BM, delayed no doubt by its shoe-box aerodynamics. The Rover's top speed went up by about five mph, to an average of 113 although the test was plagued by a surfeit of the country's most recent gales so our one-way best of 117mph was probably more representative. More to the point, the bigger engine's muscle brings a continental motorway cruising speed between 100 and 110mph well within the car's usable performance envelope.

The test car's '570Ci' specification is only recent with the emphasis on torque rather than ultimate horsepower numbers. The ease with which the box shifts down two gears from 40mph, allowing the Chevy to whisk you firmly into the middle distance, is quite exhilarating in something this size. Equally impressive is the car's ability to press on from 70mph with just the one downchange. At all times the Chevrolet unit integrates perfectly with the ZF box, shifting smoothly and accurately, and the combination manages those tricky full-power double-gear kickdowns without a neck-straining surge. Chevrolet power is generally effortless and fuss-free, although it isn't quite as silky as you might have expected from a low-revving eight-cylinder engine. It's definitely less strained at maximum rpm than the engine it replaces but the odd tingle and hum still feeds through the bodywork when you really spin it up.

The power hike is obvious and is nicely engineered but it's no more than you would expect from the insertion of extra litres. More impressive is the suspension conversion which attempts to deal with some of the RR's vintage handling characteristics. Stage one is included in the engine swap and comprises stiffer springs, uprated shock absorbers and bigger anti-

● ***Range Rovers used to have washable plastic seats and rubber on the floor but this one has more leather than a Soho sex shop. And a whopping 285bhp Chevy 5.7-litre engine***

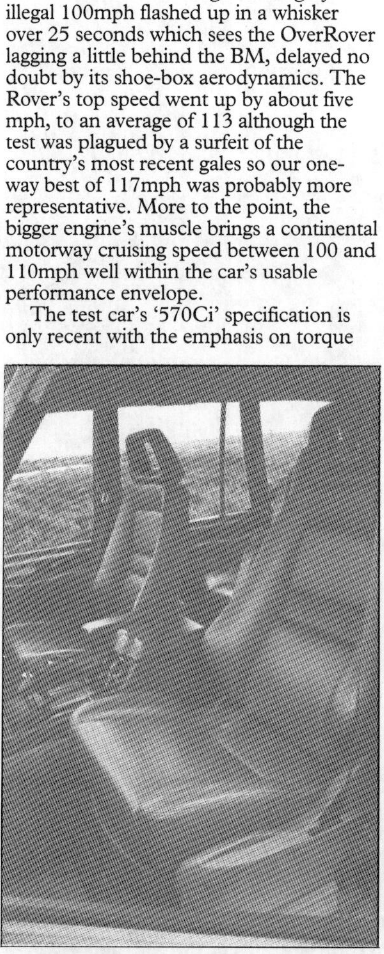

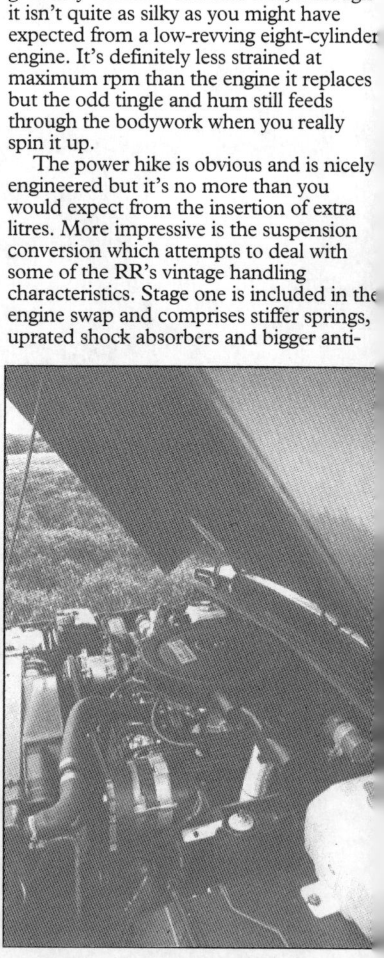

roll bars front and rear — which in the case of pre-1991 models means sorting out some roll control where before there was none at all.

There is also replacement of rubber bushing in the steering linkages which is designed to calm some of the vintage shuddering as the wheels shimmy over rough surfaces. Dedicated Rangies may miss the uncertain struggle with the wheelrim, but not me. The test car featured an extra £485 worth of modification to the steering box, which reduces the turns from 3.7 to 2.9 lock to lock, and quickens the steering in proportion. A further £550 electronically reduces the amount of power assistance in proportion to road speed. Brakes are standard Range Rover vented discs (which will be fitted to pre-1990 models as part of the conversion) and have proved perfectly adequate.

Quicker steering would be a major hazard if the Range Rover's familiar drunken body sway were not restrained, and part of the test car's £1495 stage two suspension conversion is set aside for the elimination of the squidgy rubber bushing in the links locating the rear axle beam. The rest attaches an extra pair of Boge self-levelling dampers at the ends of the rear axle rather than singly in the middle as per standard.

The benefit of all this is actually greatest when nipping and tucking through traffic or just out of the town. Quicker, more accurate steering hides the Rover's bulk extremely well while the new suspension keeps the body in check which is much more comfortable for those inside. Beyond 60mph, the beam axles start to show their vintage and on uneven roads they do steer the car — only gently, but firmly nonetheless. It's something you do get used to, but at first there's a tendency to try to correct every little deviation. After a day or so, you learn to deal only with the big ones.

The good news about the wandering unsprung weight of vintage axles however is their immense strength and simplicity, and their geometric honesty. Wheels on the end of an axle can't do much else but stay flat in contact with the ground, so the maker's intended camber, castor and toe settings remain largely unaffected by suspension movement. Generally, braking, lifting off or getting it wrong doesn't then spring any surprises. The Overfinch couples this with a general tightening of all the masses and its natural tendency is gentle understeer. More power or more speed and the nose simply runs wider, returning to line if you back off. It's all very safe and predictable, and with most of the body's slushy swaying absent, very composed and reassuring.

On the maintenance side, there aren't any wishbones to bend when you hit the verge, or universal joints to give up in the face of all that torque. Handy if the desire to be able to power the tail of your Rover into tyre-smoking oversteer exceeds the restraint imposed by your wallet. Overfinch will then be able to sell you a special transfer box for around £20,000 which is able to direct a maximum of 70 per cent of the Chevy's torque to the rear wheels rather than the standard model's equal split between front and rear axles. The extra power and additional chassis capability didn't seem to tax the brakes unduly. The pedal feels especially good with no lost movement before biting; just

● Overfinch is discreet and effective. Not only will you sit above everyone else, you'll be able to overtake them too

PERFORMANCE

OVERFINCH RANGE ROVER 570Ci

THROUGH THE GEARS (seconds)			
0-30mph	**2.4**	0-80mph	**13.7**
0-40mph	**3.7**	0-90mph	**18.4**
0-50mph	**5.4**	0-100mph	**25.2**
0-60mph	**7.5**	0-110mph	**35.2**
0-70mph	**10.2**	0-120mph	**—**

STANDING 1/4 MILE	**15.9/85 (secs/mph)**
AVERAGE TOP SPEED	**120mph (claimed)**

KICKDOWN (seconds)			
20-40mph	**2.4**	70-90mph	**8.1**
30-50mph	**3.0**	80-100mph	**11.5**
40-60mph	**3.8**	90-110mph	**16.9**
50-70mph	**4.9**	100-120mph	**—**
60-80mph	**6.2**	110-130mph	**—**
TRACK CONDITION		**Dry**	

the weight of a foot makes them work and yet they don't feel over-assisted like some.

Externally, the Overfinch conversion is subtle. There's little to tell the world that you have cash to lavish on your Range Rover, and other road-users will have to make do with the rapidly disappearing rear view and its discreet tailgate logo. The Overfinch owner will have to stump up for an 11.8mpg overall fuel consumption in return for the privilege, and will have to put up with the flimsy ill-fitting switchgear, creaking and squeaking from all the gaps in the dash and trim, and the rattles that come as standard in your average Range Rover.

The Overfinch conversion is discreet, well-developed and effective, and is not really that expensive when you look at the improvement it brings. Like most of these conversions though, it makes far more economic sense when applied to a good used — or maybe even a knackered and used — Range Rover. Overfinch will replace all the worn bits that don't come as part of the conversion, and there is a bewildering range of options to pick through — all of which can be applied to a Discovery or Defender. You can also claim back all the VAT and put it through your books as maintenance and repairs without having to argue with the authorities or your accountant. It's legal anyway, but neither would ever suspect until you pulled out to overtake. ❍

Nothing wrong with the Vogue's cabin but the new car shows improvements were possible; old dash now seems clumsy

Interior of the new car is even more sumptuous than of old. Seats are wider and more supportive; elbow room increased

WHY PAY £44,000 for a new Range Rover if you can still get the previous one, brand new, for £7500 less? The old model was the king of the off-roaders and the darling of the Sloane Square set, but the new one claims to go further still.

It's 99 percent different – longer, wider, higher, roomier, better equipped and better built, and every aluminium panel is brand new. In addition, the engines, chassis and suspension are all completely different. They want us to believe it's the next best thing to an S-class Merc, and remind us that its off-road talents remain largely untapped.

Yet Land Rover seems to be covering its back by keeping the old model on sale – wearing the label 'Classic'. However good the new car is, does it really capture the character and imagination of the old status-building original?

We took the new Range Rover out onto the streets of Hampstead with the 'Classic' to find out what the difference is. We wanted to know whether it really has been transformed as a road car, and whether the new styling has worked.

And the first question they asked us was not 'Wow, is that the new Range Rover?' or 'How much does it cost?' or 'How fast does it go?' They asked: 'What is that colour?' Land Rover's chief, having just spent a large fortune reinventing his flagship, may be disappointed.

Customers and members of the public can be forgiven for being a little confused, but what of the dealers? At the biggest Land Rover dealership in the country, Broadfields, the sales executives have had to endure a week of smoked salmon as the new car has been launched to valued RR customers.

According to general sales manager Richard Hartley, there hasn't been a single negative reaction. 'We're delighted – it's going to be very successful,' he says. 'It is more like a car now. People who are fed up after three or four Jags or Mercedes will look at it.' And is anyone going to buy the Classic as a new car? 'There's very little market, if at all, for the Classic – but there will be a second-hand market because we are going to have a lot of trade-ins,' says Hartley.

'There used to be three types of Range Rover buyer. Plant hire industrialists, farmers, and solicitors/accountants – whose customers were farmers. Nowadays our client base is wide and varied. The new RR will be used for everything because it is so quiet and comfortable.'

Does this mean it might even be taken off-road? Might it even come in muddy for a service? 'Yes,' says Hartley, 'if the Fulham farmer goes through a puddle.'

PHOTOGRAPHS BY CHARLES BEST

Vogue SE 'makes do' with 182bhp but performs well

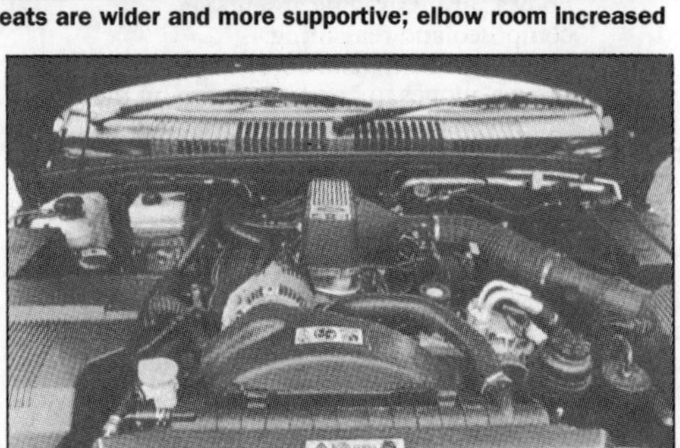

With 225bhp, the new 4.6-litre engine has effortless go

Tail of two Rovers: new car has large

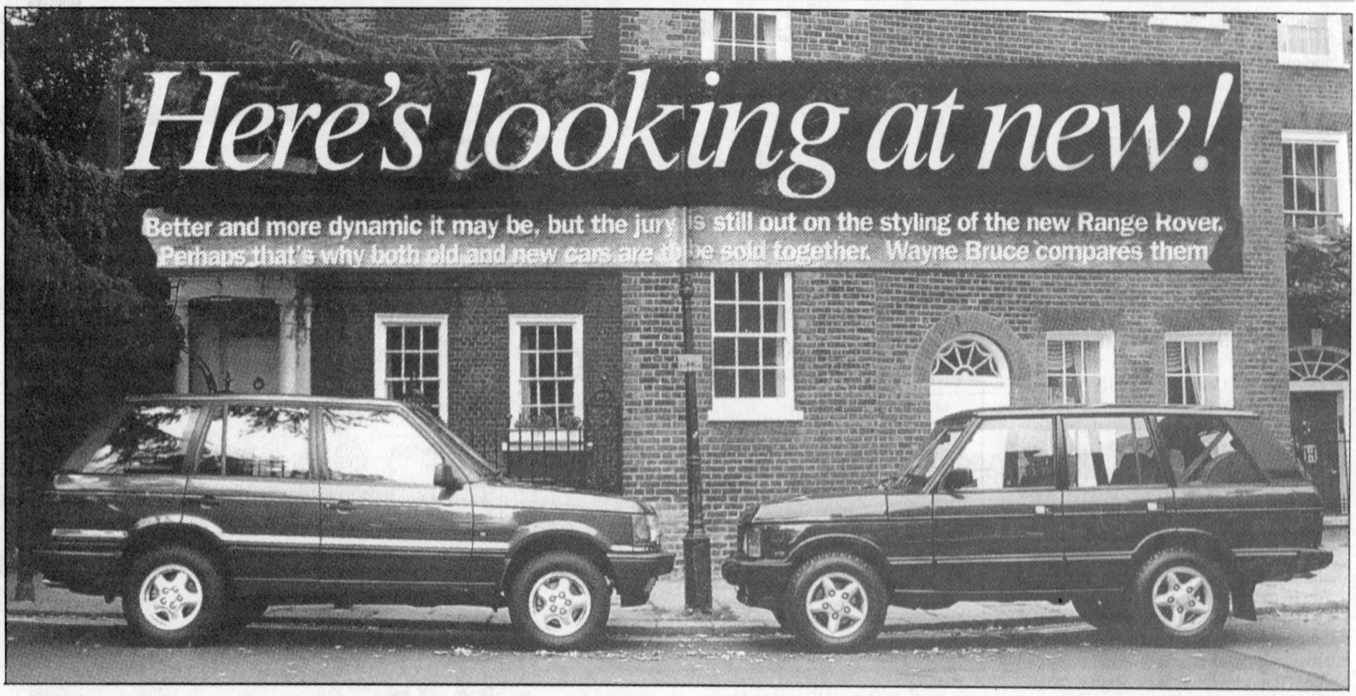

Here's looking at new!

Better and more dynamic it may be, but the jury is still out on the styling of the new Range Rover. Perhaps that's why both old and new cars are to be sold together. Wayne Bruce compares them

The new Range Rover is the first redesign for 24 years. It wasn't easy for the designers to improve on what must be the most recognisable car this side of a Roller, but some of the people we asked felt that was no excuse for making the front look like a Talbot Horizon and the rear like a Talbot Sunbeam.

'Range Rover feels, more than ever, like a luxury car on stilts'

The car has grown in every direction, but some little styling bits have been kept, so the groupies aren't offended. These include the turrets on the bonnet sides and the black rear pillars, which once represented the first use of plastic fake vinyl on a car exterior. The horsey crowd will consider it splendid that the split rear tailgate stays. To fold out the Grandstand, there is a more delicate push-push operation, and now that the handle on the window has vanished, the top part can be closed and leaned upon.

On the road, you will no longer see RRs cruising the fast lane with mud flaps – now integrated plastic – flying in the wind. Underneath, the air suspension introduced in '92 stays complete, the kneeling-camel position allowing the tight-skirted in without drama. The glorious V8 has been re-engineered into 4.0 and 4.6-litre forms. BMW's 2.5-litre turbodiesel, with German engine badging, completes the range.

Inside, the new RR retains its see-all driving position, but now it feels more like a TVR, its dash and console intruding half-way down the cabin. A telling sign of progress is the mobile-phone sized cubby-hole above the stereo. Land Rover would love to see executives ditching the massive rear cabins of S-class Mercs for the new RR; the chauffeured classes no longer have to share seat space with the rear wheelarches.

In the quest to convert saloon drivers to the 4x4 way, the Range Rover has been made to feel more than ever like a luxury car on stilts. At near three-figure speeds, its serene stablity is quite remarkable. No off-roader can compare with its chuckablity. And in a-car-for-all-reasons ability, no saloon runs it close either. It's infectious. You'll love it.

Yet the Classic is charming, too. Unlike the new one with its daunting starship cabin and bungalow-on-wheels feel, the Classic has a cabin you instantly feel at home in. The dash, updated last March, politely keeps its distance and at town speeds with the air suspension on 'motorway', the Classic can be thrown around like the new one. You can whizz a bit faster, too, because it's less heavy. Where the fun stops is on the open road, where the Classic's chassis feels loose, and the interior build quality frail. In this department, the reinvented car is as ►

t, higher bumper line, narrower roof

Growing old gracefully, but still loved

Softer, rounder, but still Range Rover

New car too 'special' to be used every day

Current Vogue owner Janet Seagrove believes her 'old' car is better looking

JANET Seagrove protests that she is not, in her words, 'your typical middle-class, stuck-up house-wife in a Range Rover.

'Our Range Rover [their third] is a working car.' Indeed it is, for when Janet isn't Range Rovering between Dulwich Waitrose and her son's school, her photographer husband comandeers the car to carry his equipment and crew to shoots all over the world. It even – shock! – goes off-road.

The Seagrove car is only a few years old, yet the mileage is nudging six figures and the interior trim is peeling off. 'We'd have to look after a new Range Rover. It's too nice inside now but it's certainly easier to drive and live with.' So how does she rate the new car? 'My Range Rover seems gutsy but they've knocked the stuffing out of the new one.' In fact, the new car is 40bhp-odd more powerful.

'Our neighbours would love it though, but for me it's too refined and ordinary to drive. It will appeal to luxury saloon-car buyers. We'll stick with the old one.' Just as well for Land Rover that Janet is not a typical Range Rover owner.

solid as the Mercedes and BMWs it aspires to. But you can pass a fist through the panel gaps on the Classic, and doors with separate window-frames shout 1970. But people still love the old dear. Everywhere we went in Hampstead, heads spun on necks, and the silent British public found their tongues. Folk kept stopping their expensive steeds and giving their opinion on a car they recognised in both guises.

But the reactions were cool. The best we got was from floppy-haired John who proclaimed the new car 'distinctive and understated.' More typical was the guy who rolled up in a cloud of expensive aftershave and a showroom-fresh Porsche 911 Cabrio – 'Hummph', was all he said. A woman walking past stopped to say 'golly, they look rather flash.' When we asked her which was the more expensive, she pointed to the Classic.

A swarm of blazered boys didn't need any asking. 'I wouldn't want to be seen in a new Range Rover,' was 13-year-old Richard's cold verdict. His older chum Patrick was less hasty – 'In time, I would learn to adapt,' he said.

The British are not great lovers of change, but in time the new Range Rover will be clogging up Hampstead streets and private school quadrangles just as the old one did. But for now, the jury remains out. The dealers are confident, and early interest has given them cause to be. As for the chap who asked for the colour of the HSE – it's called 'Avalon', which also happens to be the name of King Arthur's burial place. Two English legends together, still thriving.

RANGE ROVER 4.6 HSE — £43,950

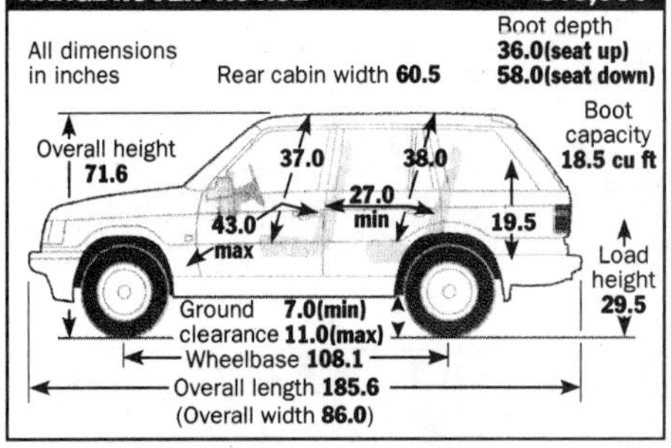

RANGE ROVER 3.9 VOGUE SE — £36,500

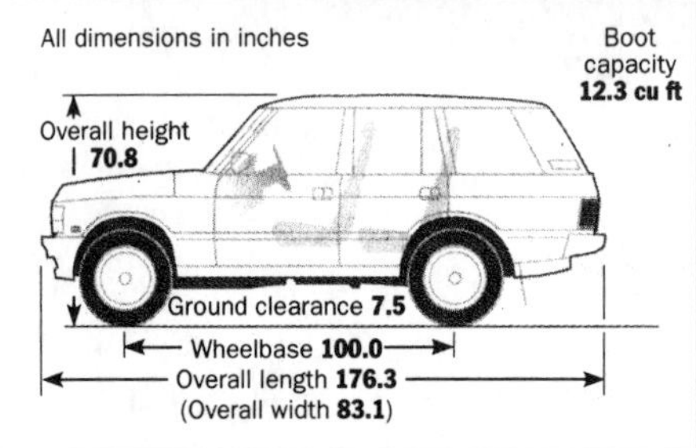

'British don't love change, but they'll get used to this one'

PERFORMANCE & SPECIFICATION

	RANGE ROVER 4.6 HSE £43,950	RANGE ROVER 3.9 VOGUE SE £36,500
PERFORMANCE		
Top speed (mph)	125	109
0-60 mph (sec)	9.3	10.4
FUEL CONSUMPTION		
Touring mpg	19.7	18.1
Gov't urban/56/75	15.2/27.2/21.0	13.4/25.9/19.8
Range at touring mpg	433	326
Tank gallons/litres	22/100	18/82
ENGINE		
Engine position	Front, longitudinal	Front, longitudinal
Cylinders/capacity	V8/4554	V8/3947
Max power, bhp/rpm	225/4750	182/4750
Max torque, lb ft/rpm	277/3000	231/3100
Bore/stroke	94/82	94/71
Valve gear	pushrod ohv	pushrod ohv
Compression ratio	9.35:1	9.35:1
Management	MPFI	MPFI
TRANSMISSION		
Type	4sp auto, 4WD	4sp auto, 4WD
Mph/1000 (top)	26.7	26.7
CHASSIS		
Front suspension	Air springs, beam axle, Panhard rod, anti-roll bar	Coil springs, beam axle, Panhard rod, anti-roll bar
Rear suspension	Air springs, beam axle, Panhard rod	Coil springs, beam axle, Panhard rod
Steering	Recirculating ball, power assisted	Worm and roller, power assisted
Turning circle, feet	39.0	36.7
Turns lock-to-lock	3.2	3.6
Brakes F/R	Vent disc/disc, ABS	Vent disc/disc, ABS
Tyres	F 235/70 R16, R 255/65 R16	205 R16
WEIGHT		
Kerb weight, kg	2220	2010
Power/weight ratio	102bhp/tonne	91bhp/tonne

Way to go: both equally at home in the city and country. Old car is about to be overtaken in almost every department by the blander, but more dynamic new car

First & Last Range Rover

Nobody had ever seen anything like the Range Rover when it appeared in 1970. In the USA there had been a few leisure 4x4s but they drove like trucks, handled like errant barges and looked like converted panel vans. The Range Rover, by contrast, was strikingly attractive, had the speed, handling and braking of a 2-litre saloon and enjoyed a supple and car-like ride on its long-travel coil springs – and yet off the road it was every bit as able as a Land Rover. It was an irresistible cocktail if you had the money: at £1,998 it cost around the same as one of Rover's big luxury saloons, with which it shared its light-alloy V8.

The run-out model, almost 25 years later, is still much admired, despite being crowded by (mostly Japanese) rivals. Land Rover moved it upmarket in the Eighties, making it compete on price and specification with conventional luxury cars. This policy worked brilliantly – Range Rover sales actually increased as a result. Despite a drop during the recent recession, 1993 sales were still higher than sales at any time in the 1970s or 1980s!

THE ORIGINAL

YVB 172H is one of a pre-production batch of Range Rovers built in the first half of 1970. Since retirement it has belonged to the well-known collection which is now

King off the road – but no mean beast on the road, either. Emphasis changed to luxury over the years, though; enamelled steel wheels (below) became elegant alloys (below right)

1970 Range Rover – first of the breed

Cost new: (prod model) £1,998, inc £469 7s 6d tax; chassis number 35500020A. YVB 172H is a pre-production Range Rover built in early 1970; used for management assessment. Owned by the Dunsfold Land Rover Trust (tel: 0483 200567).

Two-door body with horizontally-split tailgate. Metal grille with vertical bars. Painted rear quarter panels and exposed fuel filler. Enamelled Rostyle steel wheels. 3.5-litre carburettor 135bhp V8 engine, four-speed manual gearbox with integral two-speed transfer box and permanent four-wheel drive. Lockable limited-slip centre differential (now replaced by later type). Coil spring suspension. Four-wheel servo-assisted disc brakes. Manual steering. PVC interior trim, floor covering. Multiple-panel dash; small main instrument panel and one auxiliary instrument in dash centre. Three-spoke steering wheel. **Max speed:** 92mph; **0-60mph:** 13.9sec; **Fuel consumption:** 14.4mpg; **Wheelbase:** 100in; **Overall length:** 176in; **Width:** 70in; **Height:** 70in.

Top, same light-alloy V8 but enlarged to 3.9 litres for later car (left); fuel injection came in 1986. Difference on the road is telling, but the original is no slouch

called the Dunsfold Land Rover Trust, and all restoration work has been done with the utmost sympathy and concern for originality. Only the centre differential is not to original specification, because parts for the limited-slip type fitted when the vehicle was new are now impossible to obtain.

YVB looks as imposing now as it did in 1970, saved from boxiness by its large glass area and discreet body-side swage lines. There were just two large doors on these original vehicles – a four-door body would have been more expensive and taken longer to develop – and they leave the side view beautifully uncluttered. However, the detailing is crude. Those large black door pulls, the exposed fuel-filler cap, the silver-painted bumpers and the truck-like grille are all reminders that this was designed as a working vehicle which could double as a comfortable saloon on the road.

Inside is where the early Range Rover's workhorse nature becomes most obvious. Everything is functional, even minimal. This pre-production model even has a plain finish to the dashboard; production examples had leather-grain moulded in to make them look less basic! PVC coverings for seats and floor were supposed to allow owners to hose out the interior after a hard day's work, and the chunky door handles, window winders, switches and pedals were all designed to be operated by someone wearing working gloves and boots. The fact was, though, that most owners saw in this dual-purpose vehicle only the smart saloon. Before long, they started asking for more luxury and less utility.

On the road is where, back in 1970, the real revelations came. If you could live with its long-travel gear-change and the intrusive gear whine from the transfer box, this was a machine for getting places very quickly. You could cruise at 85mph or more, albeit with a lot of mechanical, tyre and wind noise; the high driving position allowed you to see over other traffic to exploit overtaking opportunities; and the Range Rover stopped securely and all-square with its big disc brakes. It could hardly have been more different from the Land Rovers which figured prominently in its ancestry.

IN BETWEEN

The Range Rover changed little during the 1970s, but as the Eighties dawned the changes came thick and fast. Four-door bodies, five-speed gearboxes, an auto option and spec upgrades answered criticisms made by customers over the years. The later 1980s then saw a diesel option, leather upholstery as an alternative to the standard cloth (PVC had gone in the 1970s), a more powerful 3.9-litre engine and a quieter, simpler transfer box. Range Rovers could also now be fitted with convenience options such as electric windows and sunroof, electrically-adjustable mirrors and seats and even air con.

Not all these goodies were standard, of course: only the most expensive Vogue SE model offered everything. Even that was topped in 1992 by the extended-wheelbase Vogue LSE, which had all the luxury equipment as standard, a 200bhp 4.2-litre engine to offset the extra weight and a unique height-adjustable air suspension which was fitted to standard-wheelbase Vogue SE models at the same time. Two years later, both

1994 Range Rover Vogue SE – last of the line

Cost new: £36,500 approx; chassis number 3MA651366. 1 HYH was new in October 1994.

Four-door body, still with horizontally-split tailgate. Plastic grille with horizontal bars; plastic front spoiler. Vinyl-covered rear quarter panels and concealed, high-mounted, fuel filler. Alloy wheels. Catalyst-equipped 3.9-litre fuel-injected engine producing 182bhp. Four-speed automatic transmission with separate two-speed transfer box and permanent four-wheel drive. Automatic locking centre differential. Height-adjustable air suspension. Four-wheel servo-assisted disc brakes with anti-lock. Power-assisted steering and traction control system. Leather upholstery with high-grade woollen carpets. Single-piece dashboard with wood veneer inserts and integrated instrument binnacle. Two-spoke steering wheel.

Max speed: 110mph; **0-60mph:** 11.3sec; **Fuel consumption:** 16mpg; **Wheelbase:** 100in; **Overall length:** 176in; **Width:** 70in; **Height:** 70in.

Stately progress: wood and leather of modern Range Rover (above) evolved from crude workaday PVC cabin, silver-painted bumpers and exposed fuel filler cap

standard- and long-wheelbase models gained extra safety features, a vastly-improved one-piece dashboard and were renamed Range Rover Classic models when the brand-new Range Rover arrived in September 1994.

THE FINAL MODEL

The 1994 Vogue SE we borrowed for this comparison is however a more direct descendant of the original Range Rover, not least because it sits on the same 100-inch wheelbase. The family likeness between the 1970 pre-production vehicle and the Vogue SE is immediately obvious, but in fact almost everything on the car has been changed. Obvious at the front are the suave horizontally-slatted plastic grille and matching chin spoiler. Smart alloys replace the steel wheels, while door and bonnet hinges have gone; under the skin the bodyshell is welded rather than bolted together, and of course there are air springs and the ABS introduced in 1989.

Inside, acres of leather are complemented by discreet figured wood; hardly suitable for hosing down. The dash, complete with airbags for both driver and passenger, is quite clearly modelled on saloon car practice and makes those used to conventional luxury saloons feel at home immediately on sitting behind the wheel.

On the road, the '94 model accelerates with even more gusto than the original Range Rover, and the anti-roll bars (fitted since 1990) allow it to be flung around corners with similar abandon despite the air suspension and the extra weight above the axle line. It's much quieter at speed, thanks to a chain-driven (instead of gear-driven) transfer box, more effective soundproofing and less aggressive road tyres. And as there are rubber air bags instead of steel springs between axle and chassis, road noise transmission is also reduced.

But above all, the '94 Range Rover is much easier to drive than the original. Turn on the air conditioning, sink back into the leather armchair and let the automatic transmission and power-assisted steering do their work. Going off road? Simple. Just slip the transfer lever into 'Low' range and let the auto gearbox, rear axle traction control and automatic locking centre differential do all the work. If you make a mistake, the ABS will help stop you scratching the high-quality paintwork, even if you do the unthinkable and brake on a muddy downhill run.

The fact that the Range Rover has remained in production for so long speaks volumes about the soundness of its basic design. But expectations change in the motoring world, and the changes which Land Rover did make over the years changed the balance of the vehicle. What had started as a workhorse which could double as a smart saloon turned into a luxury saloon which could double as a workhorse. By 1994, not many buyers of a new Range Rover wanted to sling bales of hay in the back and go trundling off across muddy fields.

However, perhaps the most impressive aspect of that change of direction is that the later, plusher vehicle lost none of the abilities of the original. Off road, with the right tyres, the 1994 Vogue SE is every bit as able as its ancestor. And on the road, it is faster, quieter and can be driven just as hard with less effort.

Yet the original 1970 Range Rover remains a mightily impressive vehicle. The fact that Land Rover actually managed to improve so little on the basic characteristics of the original Range Rover in 25 years shows just how far ahead of its time it was when new. Land Rover surely has every justification in rechristening the run-out model a Range Rover Classic. CLASSIC CARS

With thanks to Cliveden House Hotel, Cookham, Berks and Dunsfold Land Rover Trust

Major modifications, 1970-1994

1970	Range Rover introduced.
1973	First major revisions, including power-assisted steering option.
1975	Option Packs to improve equipment levels.
1979	Facelift, including black bumpers and new badges.
1980	Limited-volume Monteverdi four-door conversion.
1981	Two-door Range Rover In Vogue special edition model. Four-door production model introduced.
1982	Three-speed automatic transmission option introduced.
1983	Five-speed manual gearbox replaced four-speed.
1984	Range Rover Vogue a regular production model; new instruments, interior trim.
1986	Two-door models dropped for UK. Turbo D introduced with 2.4-litre VM engine.
1986	Fuel injection for Vogue models; new four-speed automatic; facelift with new grille, spoiler, etc.
1988	Vogue SE introduced. Chain-driven transfer box replaced gear-driven type; 3.9-litre engine for USA only.
1989	3.9-litre engine replaced 3.5-litre for all markets; 2.5-litre VM diesel replaced 2.4-litre in Turbo D; ABS introduced.
1990	Anti-roll bars introduced.
1992	Long-wheelbase Vogue LSE arrived; air suspension for top models; 200 Tdi engine replaced VM 2.5-litre.
1994	New dashboard with airbags; new R380 five-speed gearbox; improved 300 Tdi engine; SE trim and automatic options on diesels. Range Rover renamed Range Rover Classic on the introduction of the brand-new Range Rover model.

1st

RANGE ROVER CLASSIC

A CLASSIC HIT

Still a tremendous vehicle and now cheap to buy

What goes up must come down, and the once high used prices of Range Rovers have now obeyed the forces of gravity and have become eminently affordable, considering how luxurious and prestigious the package is.

Since the second-generation Range Rover appeared in 1994 values of the old model – now referred to as the Classic – have been falling steadily. This time last year the last of the Classics (a '95 M-reg) would fetch £17,500, but that's now down to £14,000-£15,000.

Of course you needn't pay anything like that for a Range Rover. A thousand quid will buy you a '70s example, but if you want something decent then £3000-£4000 gets an '88 3.5 EFi with the revised, whine-free viscous-coupling centre diff. If you want a 3.9 – and who doesn't – add £1500.

By 1994 the Range Rover was getting a bit dated, but it still provides a King Of The Road feel that few other 4x4s can offer. And while the Range Rover is a country set vehicle, it's brilliant in town thanks to top visibility and a relaxed driving feel.

If you can't cope with the 10-15mpg consumption there's the 2.4/2.5-litre turbodiesel which will give around 25mpg. You lose the V8's smoothness, and year-for-year prices are higher due to the Turbo D's scarcity. So in overall terms it's not necessarily cheaper.

What to look for

It makes sense to go for the highest equipment spec

possible, in other words the SE with its leather and air-conditioning. The ultimate Classic was the LWB LSE 4.2 but there aren't many around.

It needs to be understood that Range Rover build quality – though improved down the years – was never anything special. So be prepared to lavish constant care and attention on one.

The V8 engine can take high mileage, but may start overheating due to oilway sludge – invariably a result of insufficient oil changes. A rattling from the head means a replacement cam is needed, not a cheap job given the V8's overhead valve design.

Another expensive item that seems to go is the ECU, the engine 'brain' mounted under the driver's seat. Though simple to replace the unit is expensive to buy.

The transmission is usually bomb-proof, but on pre-'88 vehicles expect a noticeable amount of gear whine and some driveline slop every time you change gear. From the mid-'80s onwards the V8 was increasingly ordered as an automatic, so if it's a manual you want, you might have to mount a serious search for it.

Bodywork is mainly aluminium, but a Range Rover can still be a rust bucket, with the metal floor and upper and lower tailgates corroding badly. You could even see quite a lot of rust on a 1990 model.

But right now is a great time to buy a Range Rover Classic. There are plenty of good ones around at affordable prices.

Sum up: *still oozes much class, amd prices are getting lower*

SPECIFICATIONS

Built: 1970-94
Engine: 3528, 3947, 2500
Prices: £500-£15,000

On-road ★★★★☆ **Off-road** ★★★★★ **Build quality** ★★☆☆☆ **Reliability** ★★☆☆☆ **Style** ★★★★★

Second-hand Range Rovers with many extras and low mileages can be picked up for a good price

Pick of the Range

The Range Rover Classic delivers the best in on and off-road performance. Today they're an attractive prospect for the second-hand buyer, says James Ruppert

As the new Range Rover hits the £50,000 stratosphere, the legendary Classic model can be picked up for just a few thousand. And looked at in the light of a brand-new Defender, or even a slightly-shabby 110, an old Classic looks like the comfortable and highly-affordable alternative. You can still pay around £20,000 for a late model LSE (original list price £40,000), yet a tidy 1988 Efi is only going to be around £6000.

Range Rovers don't date... but they can break down.

The Classic has had its problems, but it is essentially a sound piece of kit. After 28 years in production, Range Rovers are well proven, there are plenty around to choose from and a whole support industry exists just to supply parts and services.

Providing you can afford to keep up with the heavy fuel consumption you'll get the respect of other on and off-road users, and the best dual-purpose vehicle that not-much-money will buy.

Early models and rare limited edition models are popular

THE ALTERNATIVES

V8

(1970-79)
Range Rover in its purest form: two doors, rubber mats, plastic seats, four speeds and a V8 engine. It is highly unlikely that many owners will have left them in standard trim. Modification milestones are 1973 (interior uprated, power steering became an option) and 1975, when the gentrification process began and options included headrests, tinted glass and cloth seats. Overdrive not offered until 1978.

(1979-84)
Better-built, higher equipment levels, cloth seats and standard power steering make all the difference. Four doors from 1981 and automatic gearbox available a year later. Five speeds and improved suspension make it a better drive. 'In Vogue' limited editions in 1981, 82 and 83.

(1979-85)
Basic (vinyl seats, rubber mats) two-door and van version offered to

Range Rover – important points to check

Bodywork
Most panels (except bonnet and tailgate) are made of aluminium, so they resist rust. However, they don't always fit well (especially 70s examples) and dent easily. Corrosion does occur on wing edges and door bottoms where road salt reacts with the alloy. They will also corrode where they touch the steel frame, so look at the top outer edge of the front wings. The bonnets and tailgates are notorious for disintegrating. Look for rust in the upper tailgate frame, footwells, sills and inner wheelarches. As for the frame itself, look at the door posts (A & B panels).

Interior of early early models designed to be hosed down

Engines
V8 petrol: the valvegear gets sludged up leading to camshaft wear, so listen for top end rattles. Aluminium alloy construction means the correct coolant is essential, otherwise the waterways corrode. Look for signs of overheating and rusty-brown coolant. Range Rovers used for constant towing put the V8 under stress, leading to serious bore wear, so look out for blue smoke on start-up and under acceleration.
Diesel: The original VM units could blow head gaskets quite often, so look for oil in the coolant.

Check V8 engines over carefully

Gearbox
Drive the vehicle in all its gear settings. Four-speed manuals clonk, whine and seem sloppy, but they are very durable. However, they can use a lot of oil. Check for leaks on all models and look at the levels to see if maintenance is up to scratch. The five-speeders are quieter, but listen for serious bearing whine. The box should not jump out of gear. Diff locks won't work if not used regularly. Automatics should be smooth and change down without jolts. Look at the transmission fluid, which should be red. If brown with black bits in it, failure is imminent.

Chassis/suspension
Look for off-road damage underneath. Check for leaks on gearbox casings and swivel pin housings and for corroded brake lines. Check chassis itself for signs of rot, especially the side members over the rear axle and at the rear crossmember. If the Range Rover does not sit straight, suspect worn suspension, radius arms, coil springs and worn dampers. The self-levelling strut should be able to raise a heavily-laden rear end. Test it. Unassisted steering is heavy at low speeds, while the power-assisted variety can seem vague. Wandering means worn ball joints. Look for leaks from the power-assisting pump.

Interior
Low build quality, even on the Vogue models. Combine that with a hard, Labrador/kids/luggage life and the interior can look rather sorry for itself. 1985 onward saw an improvement, but the fit and finish is still far from perfect as trim comes loose, breaks ups and the upholstery collapses. While you are surveying the damage, check all the electrical items properly. Electric windows, central locking and air conditioning are all prone to playing up.

special order only, finished in white or red paintwork.

(1984-88)
Better spec from 1984: no quarterlights, revised interior, reclining seats, electronic ignition. Vogue now production model with fuel injection. Efi engine for range in 1986, split rear seat and facelift with concealed bonnet hinges. Vogue SE specification – leather trim, air conditioning, etc – from 1988.

(1989-95)
3.9-litre engine. 1991 dual-cone synchromesh on first and second gears. 1992 Standard catalyst. 1994 new fascia, new gearbox and side impact beams. Re-badged as Classic.

Diesel
(1986-95)
Italian VM turbo diesel 2.4-litres, uprated to 2.5-litres in 1989. Available in comfortable Vogue revisions.

Limited editions such as LSEs are rarely taken off-road, so expect to find some good examples

LSE
(1992-94)
Eight-inch increase in wheelbase, 4.2-litre 200bhp version of the V8 with automatic gearbox. Revisions as second stage development vehicles (1979-84), redesignated Classic then discontinued in December 1994.

The market
A Range Rover has to be bought on condition rather than price. There is no such thing as a 'cheap' Range Rover. At the bottom of the market anything close to £1000 is virtually scrap value. Good usables are in the £2000 to £3000 category, irrespective of year.

Any Classic can be easily uprated to later specifications as sellers describe their Range Rovers as 'Vogue spec', just because there are a few bits of walnut capping in the cabin. It is also possible to make a 1970 machine look like a 1986

A big problem buying privately is that you never know how the vehicle has been used. The ideal is an urban Range Rover that has been used purely on the tarmac and serviced on the dot. Main agents prefer very recent models and charge accordingly.

Ex-police Range Rovers are very popular, having been well looked after, although the spec is basic and there are loads of holes in the facelift model with a few panel swaps, so be sure of what you are buying. The V5 registration document should help: be suspicious if it is 'lost'.

Above, check service records when buying

Right, look carefully for corrosion on bodywork

Don't be too surprised to find lumps other than the venerable V8 under the bonnet. Owners who can't take the V8's mpg and can't afford a new turbo diesel have been concocting 'strange' Rovers for years with Isuzu, Nissan and Mazda diesels, among others.

dashboard and roof. These are best bought direct at auction, but don't join the bidding frenzy and pay over the odds.

Vogues are the best buys when it comes to future resale, but they must have a nice spec, strong metallic colour and be automatics.

Range Rovers are reasonably economical to run (apart from petrol costs), but expensive to repair. Remember that and you won't buy a bad one. LRO

What you should pay

3.5-litre V8 130bhp
Four-speed. The real workhorses of the range, usually come with a tow bar and well-used load bay. Best model for those who want to 'use' their Range Rover. Avoid anything under £1000, otherwise £1500-£3000 is a good guide.

3.5-litre V8 125bhp
Four-speed or three-speed auto. More useful, more doors, less Vogue trimmings and better for it. Pay £2000-£3000.

3.5-litre V8 165bhp
Efi engine fitted to Vogue first means perkier performance. Go for the November 1986 facelifts. Best of the old models. SE-spec Vogue most in demand; for those who spend most of their lives on the tarmac a Vogue is the only way to go. Pay £4000-£7000.

3.9-litre V8 185bhp
Revised interior and starting to get complicated, especially with arrival of catalyst in 1992. Extra power useful. Prices jump for 1993-94 models, so early ones better value. Pay £8000-£19,000.

LSE 4.2-litre V8 200bhp
Eight-inch extra wheelbase. Few taken off-road, so likely to be in premium condition. More room in the back and, at £12,000 for an early high-miler, reasonable value (though you can pay up to £20,000).

Diesel 2.4-litres turbo 90bhp
Five-speed. November 1987 Vogue (From October 1989 2.5-litres, 119bhp). Frugal, but funeral pace. Bigger unit with improved mpg and more torque low down. Pay £5000-£17,000.

RANGE ROVER

PRICE RANGE £2500 (decent pre-'80 3dr 3.5 V8 four-speed); £5000 ('85 5dr 3.5 V8 5sp); £10,000 ('90 Vogue 5dr 3.9 V8); £25,000 ('95 Vogue LSE 5dr 4.2 V8 auto)
LAUNCH DATE Introduced '70; 5dr '81; auto '82; 5sp '83; Vogue '84; VM TD '86; 3.9 engine '89; air suspension, 2.5 LR TDi '92; 4.2 auto '93
PROS Superb image, safe, rugged, terrific off-road
CONS Prodigious thirst when thrashed, body and trim get tatty

There is no better second car than a mature Range Rover – a safe, relaxing way to take the children to school, surveying the traffic from a lofty height, or to carry the baggage of family life. Range Rovers' "go anywhere, pull anything" philosophy can't help but stir the adventurous spirit; they are also easy to park, and handle much better than the pundits would have you believe. Fuel consumption of 14-23mpg and above average servicing costs are countered by cheap insurance and parts that last for decades, costing less to replace than they would in many small cars. And, when you don your dinner jacket and ball gown for the grandest of grand nights out, you can sweep up in style in your £2500 Range Rover; 16 grand's worth of Mondeo won't even get a second glance.
Richard Wilsher

●Checking it over

If you follow the trend, you'll be keeping your Range Rover for years, so make sure it's fit in the first place. Tales of Hooray Henries whacking new RRs into gateposts to prove their lack of regard for dosh are not that ill-founded. And it's also quite likely to be the oldest car you will ever buy. Documentation is important, but so is the well-loved feel of a tidy interior, oil change stickers and signs of grease on the propeller shaft nipples.

The body construction is alloy and steel, bolt-together, complex and prone to corrosion. Check for rust and buckling on the inner wings under the bonnet around the battery and wash-bottle, and around the radiator for leaks.

Open the doors and check for rust in the hinge area, along the door bottoms and pillars, and check that the windows work properly. Check the rear panels for corrosion bubbles in the wheelarches, and look for rust at the top and bottom of the tailgate and above and below the split halves. Look above the rear wheels at the body mounts, and around the seatbelt anchors. Check the discs; they're a frequent MoT problem.

Put on a pair of overalls, crawl under the car (the Range Rover is one of the few cars you can inspect without lifting) and check the front swivel hubs, engine sump, steering box (£250 each), pinion seals and transmission for oil leaks, rubber suspension bushes for obvious fraying (£120 for a set plus around six hours' fitting) and driveshafts for slack joints.

Spartan interior of '80 model later transformed

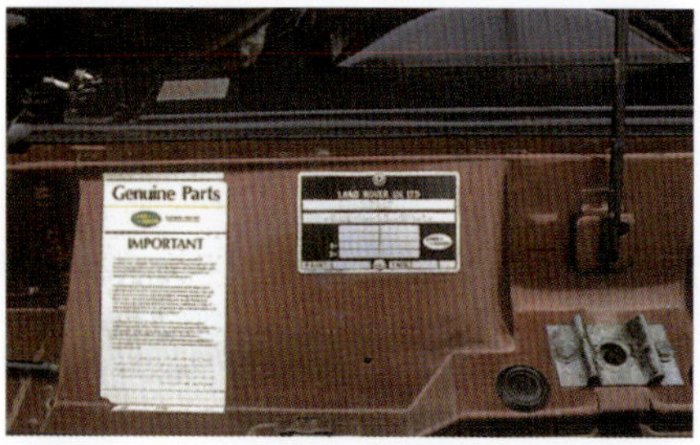

Range Rover parts prices slashed from this month

Leaking front swivel hub seal is an MoT failure

Carefully inspect the bottom of the chassis and the separate inner sills of the body mounting rails for rust – this is the killer in older cars, and a modern example used for launching boats will succumb very quickly (beware yacht club stickers).

Start the engine. A brief, heavy thumping indicates main bearing wear. This is quite common and worth a £500 reduction in the asking price, but many owners live with it for years providing the oil pressure light stays out and pressure is above 30psi hot at 50mph. Also beware tappet clatter and a rhythmic misfire. An emissions certificate is advisable; the engines are very clean if ➧

AUTOCAR
Buying USED

Rugged, stylish and surprisingly cheap to run, a secondhand Range Rover of any vintage can change your life for ever

£11,995 buys a 50,000-mile 1991 H-plate 3.9-litre Vogue with FSH – a lot of class for the money. Cheap to insure, too; under £200 fully comp for male aged 45

Tatty trim and door post rust common on early cars

sound, but poor emissions after tuning usually means valve/lifter problems and a bill of £500, or clapped injectors (£700).

Move off, checking that the gear lever does not rock backwards and forwards and that the gears engage smoothly, with a subdued whine. Early four-speeds up to '83 are very agricultural but are unburstable; the five-speeds, particularly the later short gear lever models, are much quieter and more user-friendly, but less robust. The '82-'85 three-speed and later four-speed autos make no more noise than a car transmission, while the chain-drive transfer box from '88 removed the major background whine.

Now jump on and off the throttle. Loud clanks from the gearbox mean caution, while thumping from under the rear means rear axle top joint (£50) or rare self-levelling unit (£350) failure (experts chuck the unit and fit Discovery rear springs at £75 for the pair). A specialist reconditioned manual gearbox will cost around £1000 fitted, but the transfer box bolted to the main unit will add around £600 to that if damaged. It is worth saying here that the Range Rover is a DIY delight – everything mechanical is massive, gives plenty of warning of trouble, is easy to dismantle and, by using specialist parts companies such as AEW Matlock (01629 584499), most component costs can be cut at least in half.

Finally, check the headlining, which bio-degrades, falls around your ears and costs £200 to replace.

Rugged mechanicals last decades, but body and frame rust can be costly ★★★★

Snap judgement

Taunton Land Rover's Rod Warren gave us a '91H 3.9 Vogue auto to try; probably the oldest RR Classic they would stock, although they did sell an exceptional '87 model recently. Priced at £11,995 with 50,000 miles, full service history and electric sunroof, the car impressed inside and out, being virtually unmarked apart from mobile phone mounting holes in the dashboard and a headlining bulge. With a whisker over 2000rpm and a subdued whooshing, this Range Rover accelerates purposefully like a jumbo jet. From the aircraft-like chairs the main impressions are of half-felt bumps, gentle rocking and a subdued whine, while you survey the landscape over the hedges – and have to remind yourself that you are piloting two and a half tons of metal suspended on axles two strong men can hardly lift. Everything worked well and unobtrusively – a very nice car.

Six years old, yet looks and drives better than many cars at six months ★★★★

RR always a willing workhorse, towing or toting

Test car carries scars of mobile phone mounting

£230 to fit front discs/pads comparatively cheap

Vital to check corrosion-prone inner body rails

●Parts and servicing

Taunton Land Rover's workshop controller, Tony Jacobs, says they have just heard of big cuts in Classic spares prices – brake discs down from £75 to £55 including VAT, for example. He says that major dramas were sorted decades ago, but minor updating such as new swivel lubrication goes on all the time. Air suspension needs their lap-top computer analyser in the rare event of a malfunction.

Services (6000 miles £183; 12,000-miles £275) are not cheap, but well-maintained cars never wear out. Armed with a Haynes manual and plenty of time, the DIY owner can service older cars himself, but there are at least five filters to change, and Range Rovers hold lots of fluids and have eight plugs, so consumable costs will always be more than an average saloon's.

PARTS PRICES

(Including £37.50 per hour labour and VAT)

What	How much?
Front discs and pads	£230
Rear pads	£62
Front swivel hub	£59
Exhaust	£237

Expensive routine servicing ensures very long life. Parts cheap, very durable ★★★

●Insuring it

Gossip is that Range Rovers are expensive to insure. Wrong. George and Jean have done well shopkeeping in Guildford. Now 45, with kids in college, he fancies a 1980 Range Rover to play with, tow the caravan and explore the Lakes. Though third party fire and theft is only £104, he does have a few functions to attend so a nice '90 3.9 Vogue is more appropriate, costing £191 fully comprehensive, even though downsizing the house and getting a 4.2 LSE appeals at £266 fully comp.

Nephew Colin in Barnsley has an '80 3.5 and pays £151 TPF&T, but at 30 is getting a '90 Vogue – at £338 fully comp, the insurance is half that of wife Jenny's V6 Mondeo. And if uncle George in Guildford passes on his alarmed LSE, that will cost him only £20 more.

Insurance extremely inexpensive for a such a prestige motor ★★★★★

●What they cost

Taking the five-door 3.9 Vogue as typical, the last £20,000 95M will lose about £2000 per year, meaning a 1990 example fetches about £10,000. Thereafter, it drifts down at about £1200 per year, flattening out at about £2,500 for decent 1980 3.5 four-speed cars and beyond to 1970. Really good tax exempt specimens are prized and may cost more.

Mileages of more than 100,000 are the exception, and condition is all – £6000 for a 10-year-old '87 D-plate Vogue makes some Mercedes residuals look weak.

The best post-'90 cars will be with dealers; older cars will be available from 4x4 specialists and private owners, where care is needed. So buy with your heart, but get a hard head in the shape of a Range Rover mechanic to check it out. Rare and languid turbo diesels generally fetch about 10 per cent less than the V8.

PRICES WE'VE SEEN

Model	Year	Miles	Cost	
3dr 4sp	1980	n/a	£1500	(P)
3.5 EFi 5dr	1986	n/a	£5500	(P)
Vogue EFi	1989	64K	£8250	(D)
Vogue 3.9 EFi	1990	84K	£9500	(T)
Vogue SE	1992	66K	£14,000	(P)

Key: (D) dealer, (P) private, (T) trade

Depreciates rapidly at first, then retains its value like few others ★★★★

●Living with one

The Wilsher family has had two Range Rovers over the past 16 years. The first, built in 1970 (number 154 off the line), explored Welsh mountains, towed motocross bikes, barged through drifts of sand and snow, moved us twice with a huge trailer, had its sills and floors repaired, and then went on to do the same for Mr Editor-in-Chief Cropley.

The current car, a 90,000-mile '79 model, mowed and harrowed when the tractor broke, and has been polished for special functions. The last two MoTs cost a total of £140 in specialist parts and two days' welding on sill and rear body mounts. My sister's 1984 version cost £500 at a RR dealer at MoT time for brakes, bushes, and welding after three years' use towing horse boxes; and the £400 engine top overhaul I carried out on a local '86 Vogue three years ago still sounds great as it roars past.

A Range Rover changes your life for ever – and for the better. Do it ★★★★★

WHAT HAPPENED WHEN

1970 Launched as 3dr, 3.5 carb V8, 4sp permanent 4WD with high/low ratios, diff lock
1973 Revised, PAS option
1979 Black bumper facelift
1981 3dr Vogue special edition, 5dr version
1982 3sp auto option
1983 5sp manual gearbox
1984 Vogue with new interior in full production
1985 Vogue 3.5 injection, 4sp auto
1986 3dr dropped, injection standard, horizontal bar grille
1988 Quieter chain-drive transfer box
1989 3.9 engine, anti-lock brakes
1990 Handling improvements
1992 LWB Vogue LSE, air suspension
1994 Airbags, R380 gearbox
1995 25th birthday, production ends

R380 'box prone to wear; inner wing rust spotted

ANDY CHRISTODOLO

3.9 introduced in '89, as were anti-lock brakes

Thanks to Taunton Land Rover (01823 412559) for use of cars, personnel and facilities; DM Motors, Somerset (01278 641407) for practical input; Rausch and Stallard (01702 348261) for insurance. Others in series: Peugeot 205 GTi (4.9.96), VW Golf GTi (16.10.96), Ford Mondeo (20.11.96), BMW 3-series (8.1.97)